REMOVING MOUNTAINS

Removing Mountains

· · · ·

Extracting Nature and Identity in the Appalachian Coalfields

Rebecca R. Scott

University of Minnesota Press
Minneapolis
London

QUADRANT

Quadrant, a joint initiative of the University of Minnesota Press and the Institute for Advanced Study at the University of Minnesota, provides support for interdisciplinary scholarship within a new collaborative model of research and publication.

Sponsored by the Quadrant Environment, Culture, and Sustainability Group (advisory board: Bruce Braun and Daniel J. Philippon) and the University of Minnesota's Institute on the Environment.

Quadrant is generously funded by the Andrew W. Mellon Foundation.

Chapter 3 was previously published as "Dependent Masculinity and Political Culture in Pro-Mountaintop Removal Discourse, or How I Stopped Worrying and Learned to Love the Dragline," *Feminist Studies* 33, no. 3 (Fall 2007): 484–509; reprinted by permission of the publisher, Feminist Studies, Inc.

Copyright 2010 by the Regents of the University of Minnesota

Published by the University of Minnesota Press
111 Third Avenue South, Suite 290
Minneapolis, MN 55401-2520
http://www.upress.umn.edu

Library of Congress Cataloging-in-Publication Data

Scott, Rebecca R.
Removing mountains : extracting nature and identity in the Appalachian coalfields / Rebecca R. Scott.
p. cm.
Includes bibliographical references and index.
ISBN 978-0-8166-6599-0 (hc : alk. paper) — ISBN 978-0-8166-6600-3 (pb : alk. paper)
1. Coal mines and mining—Social aspects—Appalachian Region. 2. Mountaintop removal mining—Social aspects—Appalachian Region. 3. Environmentalism—Appalachian Region. 4. Appalachians (People) 5. Appalachian Region—Social conditions. 6. Appalachian Region—Economic conditions. I. Title.
HD9547.A127S36 2010
333.8'220974—dc22
2010022014

Printed in the United States of America on acid-free paper

The University of Minnesota is an equal-opportunity educator and employer.

18 17 16 15 14   10 9 8 7 6 5 4 3 2

For Naomi and Levin.
May your future be bountiful.

Medieval West Virginia! With its tent colonies and bleak hills! With its grim men and women! When I get to the other side, I shall tell God Almighty about West Virginia.

—Mother Jones, *The Autobiography of Mother Jones*

Contents

Preface

*T*HE *NEW YORKER* RECENTLY PUBLISHED an article about the latest mass extinction event that has apparently been happening for the past fifteen thousand years or so, since humans started spreading around the globe (Kolbert 2009). The extinctions have been speeding up lately, with climate change, pollution, and increasing global connections that foster the spread of disease around the world. Amphibians that have been around since before the dinosaurs are disappearing. A virulent plague is killing the bats in New England. The article highlighted the apparently causal connection found in the fossil record: wherever humans have appeared, from Australia and New Zealand to North America, a wave of extinctions has ensued. Scientists think early human migrants might have used fire in hunting, destroying ecosystems, or simply hunted animals to death. Are humans inherently destructive then?

Clearly, humans do have a great capacity for destruction. But some human groups have achieved a sustainable balance. Maybe this is the true meaning of progress—learning to live sustainably in a place over a long period of inhabitation. There is hope for reorganizing human life in ways that can exist in a balanced relationship with the rest of life on this planet. Some ways of relating to nonhuman nature are obviously better than others, even if none is perfect. All relations to nature are entangled in social relations as well; we relate to nature in ways that are shaped by, and productive of, the ways we relate to each other. As Bruno Latour points out, the social/natural distinction is a fundamental underpinning of the specific cultural formation known as "modernity" (Latour 1993).

This book examines one instance of human–nonhuman relations, mountaintop removal coal mining in southern West Virginia. This landscape is saturated with meaning; it is constructed as a sacrifice zone at the same time that it represents a stage for the enactment of central narratives of American national identity. Mountaintop removal is, in a sense, a concrete expression of a set of cultural formations that objectify and commodify nature and people, in which the accumulation of profit is endued with moral value. Fortunately, this is not the end of the story for human beings; we have other models, other ways of being that could allow us and other species on this planet to thrive. One thing that might help in this process is the realization that we have nowhere else to go.

Broadly speaking, the environmental justice movement provides a model for acting on this knowledge. In the "power-geometry" of place and space, some human beings are much more loosely connected to a particular piece of Earth than others (Massey 1993). While some people travel more or less freely around the nation or the globe, others remain in place. Even when their neighborhoods have been used as toxic waste dumps, when the landscape has been irradiated by nuclear weapons testing, or when the mountains have been made low, some people stay, either because they lack the means to move, because place represents a connection to family and community, or because of an inseverable connection to the land. Sometimes people are forced to leave and form new connections in new places. Therefore, the movement is committed to building difficult coalitions, living with tension, and accepting the incompletion that comes with process. Environmental justice means working with the materials available, all with an eye on the long term.

This book involves the participation of people with diverse political perspectives. I hope it is clear in my analysis that even when I discuss processes that I believe are harmful to people and the environment I do not aim this criticism at individuals but at the discourses and structures that shape the frameworks we all inhabit. These discourses and structures do not just shape our individual identities and our interpretations of events; they also help determine our communal relationship to nonhuman nature. Modernity's powerful universals (for instance, that great empty grid of the globe to explore and fill in) underlie the abstract notion of space that makes some places seem disposable (Thongchai 1994). Science fiction and the space program carry on this abstraction as we moderns desperately seek a new frontier. But universals take place through the gritty concreteness of encounters across difference—through the violence of imperialism, for example, or through the free-for-all destruction of the frontier (Tsing 2005). Universals such as the liberal ideal of equality do not exist except in their concrete expressions in inequalities of nationality and citizenship (Kazanjian 2003). These abstractions carry affective connotations, feelings of safety, morality, and pleasure; the people cannot be evacuated. Where some find their identities rooted in a particular landscape, others identify with mobility. What new universals, rooted in place but global in their impact, could emerge from a reckoning, on the part of those of us less connected to place, that this is all we have?

Acknowledgments

I OWE A GREAT DEBT to all the people who took time out of their busy lives to participate in this project. I also gratefully acknowledge the institutional support for my research from the University of Missouri Research Board and the University of Missouri, Columbia, Research Council; the University of California; the University of California, Santa Cruz, feminist studies department; and the University of California, Santa Cruz, sociology department. I thank my wonderful adviser and mentors, Anna Tsing, Herman Gray, and Melanie Dupuis.

For their friendship, support, and inspiration, I thank my colleagues and teachers Barbara Barnes, Tanya McNeill, Sarita Gaytan, Elizabeth Bennett, Celine-Marie Pascale, Kim Tallbear, Andy Smith, Macarena Gómez-Barris, Dana Takagi, Andy Szasz, Julie Bettie, Jim Clifford, John Brown Childs, Craig Reinarman, Donna Haraway, Pam Roby, Candace West, Pat Zavella, Marcia Millman, Ben Crow, and Wally Goldfrank. Thanks to my colleagues at the University of Missouri, who made me feel welcome in a new place: Wayne Brekhus, Ed Brent, Eric Brown, Brian Colwell, Rebecca Dingo, John Galliher, Jay Gubrium, Peter Hall, Joan Hermsen, Victoria Johnson, Byung-Soo Kim, Jackie Litt, Clarence Lo, Rebecca Meisenbach, Tola Pearce, Amit Prasad, Srirupa Prasad, and Takeshi Wada. A special thanks to Dave Brunsma and Mary Jo Nietz for their help.

Thanks to my graduate students in the Culture, Difference, and Inequality seminar for helping me work through these ideas. Through numerous talks, presentations, and revisions, I have been fortunate to receive feedback from Rebecca Klatch, Michael Messner, Judith Kegan Gardiner, Valerie Kuletz, Ian Boal, Avery Gordon, Barbara Ellen Smith, James Scott, Kathryn Dudley, Kai Erikson, and Kathleen Stewart. Thanks also to anonymous reviewers at the University of Minnesota Press and at *Feminist Studies* for their comments.

Jesse Van Gerven and Mark Speckman provided terrific research assistance. Thanks to Debbie Friedrich and Mary Oakes for their assistance. A big thanks to Sheiva Rezvani for pointing me to Track 16 Gallery in Santa Monica, California.

Thanks to Jason Weidemann for his belief in this book and to Danielle Kasprzak for her guidance. I thank Terry Burdette at the Madison, West Virginia, post of the Veterans of Foreign Wars for helping me obtain a copy of *The Price of Freedom*. Warm thanks to Patty Viera and Ken Light for granting permission to reprint the illustrations in the book.

I received quite a bit of practical support during my research, especially from Phyllis Bunch, April and Jake Bunch, Adoree Myers, Troy Valenzuela, and Erin Baker. Thanks to my family, especially Betsy Scott and Brian Thornily, David Scott, Jim Scott, and Lynne and Arnie Garson. I would like to thank my mother, Elizabeth Harvey Scott, who is with me in spirit. Finally, love and gratitude to Scott, Naomi, and Levin.

The Logic of Extraction

*I ain't going to move, but somebody asked me, "If you had to move,
where would you go?" So I said, the first thing I'd do is go and get me a
lump of coal, and I'd go to state to state, and [ask]: "You know what
this is?" And if they said no, that's where I [would] want to live.*

—Gus, white resident of Blair, West Virginia

SOMETHING IRREVOCABLE IS HAPPENING to the land in southern West
Virginia. The coal-mining technique known as mountaintop removal
(MTR) is permanently altering the topography of the Appalachian Mountains. Even though this process is quite literally explosive (workers use large
amounts of underground explosives to loosen the rocks and soil above the
coal seam and huge earthmoving machines to remove this overburden and
mine the coal), the process is happening quietly, it seems, behind people's
backs. Usually mountaintop mines are hidden just behind the ridge visible
from the road; often an educated eye can detect a mountaintop mine signaled by sparse tree growth at the top of a ridge where normally there
would be thick forest. This type of sparse tree growth lines the major interstate highway through the coalfields, but MTR is all but invisible to the
casual observer. It becomes harder to ignore when you get off the highway.
Flattops and valley fills appear more and more often; they are increasingly
visible from population centers such as the city of Logan. MTR is rapidly
becoming part of the everyday landscape, making its drastic alterations of
this landscape seem ordinary.

In no time at all, thousands more acres of the Appalachian Mountains
will be dismantled and reclaimed as flattops with rolling grasslands and
scrubby shrublands, a brand-new ecosystem to replace the mixed hardwood forest. The characteristic experience of modernity—for Marx, the
feeling that "all that is solid melts into air"—is apparent here in the speedy
reconfiguration of high places made low (Marx 1848, 476). Once a mountain
disappears, how do we know it was ever really there? It becomes a ghost,

nearly possible to ignore. The repetitive redoings of modernity, the planned obsolescence, the constant remaking of the dualism of future and past make it hard to see what was there only a few moments before. What we see appears natural, as if it had been there always. Especially now, this is a painful prospect, as we look into a future very different from the now. Fossil fuels threaten much of the world as we know it; coal-burning power plants are certainly one of the biggest contributors to global climate change. Will the world as we are remaking it seem natural to our children?

In a political climate where the causes of climate change remain subject to debate, the practice of MTR is a site of convergence for environmental politics, economic crisis, and American national identity. On this contested terrain, the moral question of what kind of people "we" are gets played out in different ways of relating to nature. Not only questions of work, independence, and unemployment but everyday practices and cultural forms come together in hegemonic definitions of what is "real" and possible (Rowbotham 1974). In a site where the links between "analytic subject, concept and world" are so familiar and deadening—where, in other words, the answers are always already understood—it is worth slowing down "the quick jump to representational thinking" that obscures the way that the present is made up of "heterogenous and noncoherent singularities" (K. Stewart 2007, 4). The structures of feeling that animate MTR exist in the frictions of everyday encounters where mining "literally takes place" (K. Stewart 2007, 3; Tsing 2005; R. Williams 1977).

The Friends of the Mountains is a network of local, regional, and national environmental organizations and activists who are working hard to protect the mountains and mountain communities using every tool at their disposal: protests, music, rallies, lectures, legislation, the National Register of Historic Places, lawsuits, performance art, film, and the Internet, among other things. Though one of their central concerns is the survival of Appalachian communities and ecologies, these groups also address a wider set of environmental justice concerns, including toxic waste, air and water quality, and global climate change. As the Bush administration prepared to leave office in 2008, activists were able to draw more public attention to the effects of an energy system dependent on coal-burning power plants. Under the leadership of President Barack Obama, these causes can potentially benefit from a growing public interest in environmentally sustainable jobs, green infrastructure, and renewable energy. Nationally, organizations like Power Shift, Repower America, We Can Solve It, and

Kilowatt Ours reflect a growing common concern with these previously marginalized issues.[1] Today anti-MTR organizations are growing and are more prominent than ever in the coalfields and around West Virginia. Public pressure on the coal industry has been intensifying since the Sago mine disaster, in which twelve miners were killed in January 2006. The connections between the sacrifices of miners and more widely shared environmental problems became clearer yet again in December 2008 with the collapse of a coal ash containment pond at a Tennessee Valley Authority power plant in eastern Tennessee that covered hundreds of acres with toxic coal sludge (Sledge and Paine 2008). Coalfield activists are increasingly focusing on the health effects of mining and coal processing. Selenium from a large MTR mine in Boone County, West Virginia, has been linked to deformities in fish in a nearby river (Ward 2008c). In addition, the residue of coal processing and coal slurry impoundments in groundwater have toxic effects on human health, especially in communities that use well water. The toxins and heavy metals found in coal slurry can erode the enamel from children's teeth and are thought to contribute to kidney stones, gallbladder disease, and cancer (Fleming 2009; Stout and Papillo 2004).

The activism of the Friends of the Mountains has elicited a range of reactions from their neighbors in the coalfields and around West Virginia. In 2000, Arch Coal closed the enormous Daltex mine on Blair Mountain after a court ruling by Judge Charles Hayden called for enforcement of the Clean Water Act. Many coalfield residents were angry and fearful; coal companies threatened widespread job losses, even the end of mining, if the ruling held. It is widely believed that Al Gore sealed his fate in West Virginia by publicly opposing MTR (Goodell 2006). By 2004, the extreme polarity that marked the community in 2000 seemed to have receded. Activists were still continually threatened and harassed by coal companies and their proxies, but with a sympathetic federal administration in power, the coal industry's hysteria seemed to have calmed. A 2004 Democratic poll found that 56 percent of West Virginians opposed MTR, yet portraying John Kerry as an anticoal candidate proved successful for George W. Bush in that year's election (Ward 2008a).

The industry has reacted to the public's growing environmental consciousness with a campaign promoting the idea of clean coal. Clean coal refers to a constellation of conjectural, emergent, and actual techniques for mitigating the effects of coal-burning power plants on the environment, including carbon sequestration, coal gasification, and scrubbers installed on power plants to reduce smog and acid rain. All major presidential candidates in the 2008 election used the phrase "clean coal" in speeches, despite the fact that much of the technology is still theoretical, prohibitively expensive, and its effectiveness contested. More to the point for the purposes of this book, the harm caused by coal goes beyond its climate impacts and negative effects on air quality. Conditions in coal mining regions and around coal-burning power plants make the idea of clean coal seem oxymoronic (Wald 2007, 2008). Nonetheless, the industry tends to argue that alternative energy solutions such as solar, wind, or increased efficiency are unrealistic. Mining equipment company Walker Cat's billboards in the coalfields proclaim "Coal. Yes, Coal," a position in line with the view of the world's largest coal company, Peabody Energy (Byrnes and Aston 2007). For every bumper sticker proclaiming "I Love Mountains and I Vote," there seems to be one that counters, "I Love Coal." Indeed, in many conversations I had in the coalfields, environmental activists were portrayed as radicals; many people try to represent a fair and balanced perspective in which MTR and coal-burning power plants are acceptable if "done correctly."

The construction of this "fair and balanced" perspective brings up the questions at the center of this book. The activists who are fighting MTR argue that the technique is merely the latest and worst instance of the coal industry's hyperexploitation of the coalfields. The environmental devastation of mountaintop mining is something that almost no one apart from the most enthusiastic industry apologist can deny; it is inherently destructive. Hundreds of thousands of acres of hardwood forests and their ancient geological firmament are being dismantled. Communities are being destroyed. Local specificities and ways of life that have developed over time are being threatened, yet MTR is somehow normalized in the eyes of many, both inside and out of the coalfields.

At the same time, MTR is contributing to a shift in the terms of employment in coal mining. Mountaintop mines can produce as much coal with fifty workers as three times that number can produce working underground. Together with the mechanization of underground mining, MTR has contributed to the decline in coal employment levels from around fifty-five thousand in 1975 to around twenty thousand today, with coal production levels as high as they've ever been (West Virginia Coal Association 2008). Along with the decline in the number of workers has come a decline in union membership; nationwide, just under 30 percent of coal miners are unionized (U.S. Department of Labor 2006; Malloy 2010). The actively antiunion practices of Massey Energy Company have assisted this decline. Many coalfield residents and union supporters say Massey set out deliberately to break the United Mine Workers of America; the company's decentralized organizational structure based on subcontracting was instrumental in transforming the terms of labor politics in southern West Virginia (Brisbin 2002). Currently, Massey Energy is the biggest coal producer in the state, and fewer than 2 percent of its workers are unionized (Massey Energy Company 2008). As the effects of MTR and increasingly mechanized coal mining on the labor market and the environment become ever more obvious, potential collaborations are emerging between labor activists and environmentalists (Ward 2008e).

One reason for these emergent coalitions is that Massey Energy Company is becoming a greater and greater target for all sorts of social justice activism. Don Blankenship, Massey's CEO, attracted national attention due to his financial involvement in the campaign to defeat West Virginia Supreme Court Justice Warren McGraw in 2004 and his ties to other justices on the court, such as Brent Benjamin, who defeated McGraw, and Elliot Maynard, who was photographed vacationing with Blankenship in Monaco (Liptak 2008, 2009). The vacation controversy landed the CEO in the news again when he allegedly manhandled and threatened to shoot a reporter from ABC News (J. Anderson 2008). Blankenship's strong-arm tactics and involvement in multiple lawsuits have even sparked rebellion among the company's stockholders and board of directors (Schnayerson 2008, 290).

Activists have also been successful in increasing national attention and outrage over MTR and Massey's involvement in the practice. In early 2009, Santa Clara University, a Jesuit university in California, divested from Massey Energy, saying the practice of MTR was incompatible with the institution's Catholic ethics (Barrick 2009). Anti-MTR activists were prominent at the largest anti–global climate change action in U.S. history, Power Shift 2009, in Washington, D.C. However, activists are also running into new conflicts that threaten previously easy coalitions; environmentalists from the eastern highlands of West Virginia are now fighting the installation of wind farms on their own beloved mountains.[2] These examples of emergent and threatened alliances illustrate both the instability of seemingly self-evident identities and how their embodied meanings help structure our perceptions of place, the environment, and morality (Sandoval 2000; A. Smith 2008; Tsing 2005). The banality of domination through gender, race, and class is expressed in a NIMBYism that continues to construct environmental risk in terms of "us" and "them."

Theorizing the Culture of Extraction

Given the historical pattern of boom and bust in the coal industry, one might think there was still a coal bust in West Virginia. Schools are underfunded, the downtown areas of Logan, Williamson, and Madison are largely empty, and poverty rates are high. But the coal industry is actually experiencing a boom; coalfield residents just are not getting to share in it this time. While intensive mining techniques have increased the environmental impact of coal mining on local communities, in the form of subsidence from underground mining, cracked and poisoned wells, increased flooding, dust and noise from working MTR mines, as well as the permanent transformation of the landscape, these communities are receiving a smaller and smaller share of the benefits.

Why then do some coalfield residents tolerate, and occasionally even fight for, the coal industry? In light of coal's long-term effects on the health and economy of the coalfields, what sense can we make of people's self-proclaimed love of coal? Like oil, coal is a "mythic commodity," fetishized in coalfield culture in the form of coal houses, coal jewelry, and figurines carved from coal (Watts 2001).[3] One thing love stories teach is that "something does not have to feel good viscerally for it to be a pleasure" (Berlant 2008, 14). To approach this question from a fresh perspective, it is necessary to take a step back. Calling the area "the coalfields" is convenient, but it abbreviates the struggles that have developed historically over the meaning of the place, which today are at the heart of the cultural politics of MTR. MTR is an *American* problem, and this book examines it from the perspective of American national culture. It would seem to go without saying that the southern West Virginia coalfields are a place in Appalachia, but noting this placement is important to the questions addressed here. The place is contradictorily located both at the heart of the national industrial market economy and within a marginalized mountain rurality. The complicated cultural politics of MTR are related to the productive articulations between these identities.

For most of the history of the coal industry, the most overtly political tension has been in the relationship between workers and companies. The creative tensions of this struggle contributed to a rich coal culture in the coal camps and mountain hollows but at the same time helped motivate the development of mining technology requiring less and less labor. With a decreasing membership, and less political clout, the union has become less of an oppositional voice to the industry, with some important exceptions, especially around worker safety. But the union is structurally and politically dependent on the existence of the industry. It has sided with the industry against environmental activists, although some union miners oppose MTR and advocate a renewed focus on underground mining (Fox 1999, 180; Steele et al. 2007).

However, there is another side to the politics of coal mining. The industry's dominance in property ownership and politics has defined and constructed the area as a coalfield, but the place is not totally determined by coal. Every child in West Virginia learns in school about how corporations and speculators bought up most of the coal-bearing land and mineral rights in the late nineteenth century (Caudill 1962, 70–76; Eller 1982, 72; Gaventa 1980, 53). This is the original sin of the coal industry, the moment in which the place became a coalfield, which, no matter how normalized in national discourses of private property and the free market, continues to mark a disconnect between incompatible versions of place. With the increasing use of surface mining, the contradictions between place-based mountain culture and the requirements of the coal industry have become even more obvious. These contradictions reflect deep-seated conflicts over the human relationship to nature, and the determination of rights to land, that are integral to U.S. history as well as present-day environmental politics (LaDuke 1999, 2005; A. Smith 2005).

In other words, the politics of MTR are not simply about the conflict between workers and employers over the conditions of production but are also about the meaning of locality and difference in American modernity. The groups coalescing on the environmental side are organized around a notion of Appalachia as a meaningful locality, with its own identity and importance within the nation. The conceptions of Appalachian cultural difference and environmental justice that are articulated by the anti-MTR movement are focused on usufruct rights to land and forest products, an assumption of a human right to health and subsistence, and other non-hegemonic values. Industry supporters focus instead on their place in the economy as a form of abstract locality, claiming as principle the identity of national citizen and referencing Appalachian cultural specificity primarily as a commodity in the national market. This view conceptualizes landscape essentially as property, a form of abstract space where the commodity can be cognitively separated from the human community and natural environment (Lefebvre 1991). These conflicts play out in complex ways on the terrain of gender formations, culturally embedded racial logics, class identity and habitus, and the human relationship to nature.

The developing environmental crisis in industrial modernity increasingly breaks down barriers between culturally distinct objects of study. The amorphous and border-crossing character of environmental risk has perhaps fundamentally changed the way society works (Beck 1992, 1999; Giddens 1990). The growing awareness of the unsustainability of our everyday routines has brought about a sense of disquiet, or "ontological insecurity," among environmentalists (Giddens 1984, 62). This collapsing of the bracketed concerns of modern life foregrounds the messy entanglements of material arrangements and cultural forms. For example, something as ordinary as wearing shoes affects the human perception of the environment; this separation of feet from ground reflects and reinforces the Western dualisms of mind/body and humanity/nature. Ingold theorizes that these kinds of cultural variations in human embodiment are related to differences in the human relationship with nature (Ingold 2004). This resonates with Bourdieu's concept of habitus, or the way that human groups express cultural categories in space and through the body. The overlapping, redundant repetition of similar kinds and oppositions naturalizes cultural structures (Bourdieu 1998, 2001). Thus the habit of viewing the Earth as property is related to other forms of domination, such as the domestication of animals, which Ingold compares to human slavery (Ingold 2000, 74). Indeed, as he points out, this comparison shocks those whose sensibility is shaped by the binary human/nature, but for many human groups an intersubjective relationship between humans and nonhumans (based on trust, and its "underside," anxiety) is common sense (Ingold 2000, 75; Deloria 1999; Descola 1994; Kuletz 1998; LaDuke 1999). This most basic element of culture, which relates to a question of primary importance—how we deal with our "corporeal vulnerability" (Thrift 2008, 242)—brings all sorts of divergently categorized and unremarkable things together in the "onflow" of human social life: industrial pollution and feeding the kids, family-wage jobs in heavy industry and the spatial organization of our neighborhoods, the embodied sensations of being indoors and resource wars across the globe, and so on (Pred in Thrift 2008, 5).

Nonetheless, the human/nature dichotomy remains in force, and almost every conversation about MTR evokes the aphorism that coalfield communities are simply pursuing their economic interests at the expense of the environment. Coal mining is a "worse-case scenario" of the jobs-versus-the-environment formulation (Goodstein 1999, 108).[4] Masculine-identified jobs in heavy industry or natural resource extraction have long been considered real work and in both left and right politics are understood to be the backbone of the economy, where, borrowing from Locke, wealth and property are produced through the transformation and "improvement" of nature (Locke 1690). In the United States, these jobs have been historically considered the territory of white men; this is one of the key sites in so-called reverse discrimination, when due to affirmative action a white woman or person of color is perceived as taking a white man's job (Mukherjee 2005, 15–16). This articulation has its roots in the genealogies of American racial and gender formations. From the colonial period, and through the conquest of the West, the labor of Native Americans has been erased in national discourse, particularly the labor of Native American women, via the identification of Native Americans and their labor with nature. Non-European forms of labor did not produce private property and capital, which were teleologically defined as signs of progress (LaDuke 2005, 118; Cronon 1983; Harris 1993). Similarly, unequal regional geographies of development, along with the construction of a gender and racially segregated working class, cemented the identification of progress and independence with industrial production and white men (Roediger 1991). Only through those kinds of labor identified as modern and articulated with masculinity and whiteness could development occur and the nation progress.

These mythological articulations relating white men, American territory, and the exploitation of nature can be deconstructed using the methodology of the oppressed (Barthes 1976; Sandoval 2000). This refers to the linkages that cross "scattered distances and central spaces" and negotiate "complex crevices and openings" in a destabilization of colonizing knowledge (Sandoval 2000, 29). For instance, Stoler has examined the articulations between national identity, geography and colonial domination (Stoler 1995, 2002). She analyzes the ways in which the tropical climate was understood to endanger Dutch colonists in Indonesia, and how a geographically specific education in Holland was seen as necessary for developing a true European character, with its capacity for governance. In a related vein, Slotkin sees the violence of the American frontier as essential to the development of the American national character, which for him, pace Frederick Jackson Turner, leads to "a pyramid of death-racked, weather-browned, rain-rotted skulls, to signify our passage through the land" (Slotkin 1973, 565).

However, the jobs-versus-the-environment aphorism has become common sense and reflects the well-worn articulations between a particular conception of the human relationship to nature and a unilinear notion of progress and development. For some, it is tempting, if discouraging, to see this neocolonial capitalist logic as terminally hegemonic. This is the position of DeLillo's contemporary white suburbanite, wandering in a brightly lit, confusing supermarket: *"But in the end it doesn't matter what they see or think they see.* The terminals can *decode every item, infallibly"* (DeLillo, *White Noise,* quoted in A. Gordon 1997, 15). In the glare of these supermarket lights, there is no space for failure, for breakage, or for ghosts. However, the space of the coalfields exceeds this just-so story. As Gordon puts it, "If we want to study social life well," if we want to change it, "we must learn how to identify hauntings and reckon with ghosts" (23). These ghosts are the sudden recognitions, the uncanny resemblances, the incongruities that beg us to notice the frictions that characterize the workings of power.

The coalfields are haunted by the plaintive voice of a hillbilly dirge for the mountains, as well as by ghostly traces like an arrowhead found on the ground and secreted beneath a coal-company uniform, a small talisman with the power to stop a mine, a relic of a territorial past that stubbornly persists. The place is truly haunted, not only by the history of coal-mining violence and oppression, or by the echoing injuries of long-term marginalization, but perhaps most importantly by the traces of other ways of relating to the land. "Trace is a material-less, strictly speaking, nonexistent condition which . . . is nonetheless necessary for and anterior to any production of meaning" (Kipnis in Derrida and Eisenman 1997, 143). Like the rest of the United States, the coalfields are marked by traces of Native American inhabitation, by the presence and absence of (for example) Pocahontas, Aracoma, Mingo, and Logan, which echo through place-names, local legends, and company names. Like many other Americans, many Appalachians see themselves as the inheritors of Native American traditions, perhaps with Native practices and images standing in for the ancient traditions of Europe (Kazanjian 2003, 11). Yet these ghosts persist in other ways as well. As Native activist Thunder Hawk remarks, discussing potential environmental justice coalitions, "There are some [white people] that are still land-based people, and those are the only ones you want to work with anyway" (quoted in A. Smith 2008, 202). The dichotomous logic of jobs/environment falls apart here, with this notion of "land-based people."[5]

The environment evoked in jobs versus the environment is one of wilderness protection—snail darters and spotted owls—that is easily marginalized in national discourses that naturalize wages as livelihood and reproduce a dualistic opposition between people and the natural world. Like the Native horticultural practices that created the abundance that astonished the first European settlers in what was to become New England, the basic interdependency of life is rendered invisible in a worldview that credits the free market with more productive power than Mother Nature. Ecosystems were first discursively dismantled in the form of lists of commodities, then forests, animals, and prairies were actually reduced to timber, meat, and grain (Cronon 1983, 1991). When nature is reduced to a set of false commodities, nonquantifiable values, such as the mutually generative relationships between species, are elided. A commodity, strictly defined, is an object created by human labor to be sold on the market. Natural resources, land, and other naturally occurring useful things are false commodities because they are not made by humans but appropriated and sold on the market. This leads to the "second contradiction" of capitalism, that as capital expands, it destroys its own natural foundation (O'Connor 1994; Polanyi 1944).

There is a well-known explanation for this problem: people in the coalfields are willing to sacrifice their environment for jobs because the area is dependent on coal mining.[6] By arguing that there is more to the story than this, I am not discounting the importance of economic forces in the coalfields. However, these stock phrases explain away the frictions that make capital accumulation, and in this case, resource extraction, possible. Friction is a metaphor for the "effects of encounters across difference" that actually make up global economic flows and hegemony (Tsing 2005, 6). Seemingly transparent commonsense explanations smoothly bypass the inner workings of an economy based on inequalities and cultural niches located in space (Tsing 2005). The first step in stretching out and exposing this supply chain built on inequalities of region, race, gender, and cultural citizenship is to assume that the activists are not the only interesting or agentic people in the story.

The coal industry's interests—to get the coal out of the mountains as quickly as possible, with as few workers as possible, regardless of what they leave behind—do not coincide with local interests in cultural, economic, and environmental sustainability, and most coalfield residents realize this. Indeed, some expect the coal to run out within fifty years (King 2009). Meanwhile, MTR is leaving the coalfields with a greatly simplified, barren, and toxic landscape. And there is the notorious intimidation of coalfield residents by the coal industry, a pattern that dates back as far as the existence of the industry. Today paranoia permeates the coalfields. An unknown vehicle parked near a gathering in a park creates suspicion. An activist's children find their dog, shot dead, at the bus stop. The sudden death of another activist elicits fear and concern. The historical pattern of violence against workers, activists, and their families has contributed to the construction and enforcement of limited economic choices that are reflected in the jobs-versus-the-environment dogma. Past violence and its present-day versions haunt moral judgments and affective notions of safety.

In the national public discourse, the question of how to explain what is happening to Appalachia reverts quickly into discourses of Appalachian difference. The question is tired and most often ignored. Occasionally, however, it rises to the surface, sometimes in academia or popular culture, sometimes in the national news or in casual conversations. Appalachian cultural marginalization almost always predetermines the shape of these conversations, which typically construct ordinary coalfield residents as helpless, passive, or ignorant. One thing these habits of marginalization do is diminish the "complex personhood" of the ordinary, unremarkable coalfield residents (A. Gordon 1997, 4). That is why the activists are not the main focus of this book; that work is being done elsewhere (Barry 2008; Loeb 2007; Montrie 2003).

The explanations that I have found unsatisfying, which reduce subjectivity to activism, or interests to economics, tend to ignore the cultural basis of subjectivity, place, and the human relation with nature. Both Marxian and liberal theories of the subject neglect the importance of place, race, gender, and citizenship status in the process of labor subjectification (Tsing 2005). Power is not simply repressive; rather, power is generative and creative, of subject positions, identities, desires, pleasures, and resistance (Foucault 1975, 1976). Since the beginning of the coal industry, it has been necessary to force or otherwise convince people to do improbable things like spend most of their lives laboring in a dark, wet hole underground, risking their lives for their jobs. With MTR mining the improbable task for miners is to dismantle the mountains where they live. As Tsing puts it regarding a different sacrifice zone, "Resources are made by 'resourcefulness'... [the] activity of the frontier is to make human subjects as well as natural objects" (Tsing 2005, 30). In the coalfields, for example, there is the national romance of the miner, the charisma of the power to move mountains, and the incipient nostalgia of a disappearing Appalachian culture. These sentimental stories are "juxtapolitical"; they represent the vulnerability and sacrifices of working people in a romantic story of seemingly universal human suffering that calls for compassion, not change (Berlant 2008, 2). These normative fictions provide an assurance to subjects that there is "a *whatever* point for the reproduction of a predictable life" (266). A 2008 billboard advertising jobs with International Coal Group, located along I-64 between Beckley and Charleston, exemplified this affective power. It featured a family—husband in miner's work clothes, with wife and children—a picture of happy heterodomesticity, and read, "Building strong families close to home."[7]

These affectively charged and precognitive elements of what mining means are naturalized in the sense of inevitability that characterizes the coal industry's actions in the region. This sense of inevitability prevents any real conversation or deliberation that would allow the inhabitants of the region to have a say in the future of the land; the many activists who oppose the way things are going are repeatedly marginalized in public discourse. With the hope of disrupting this sense of inevitability, this book investigates the coproduction of economic and cultural forms, affects, and traces. It therefore represents a partial genealogy of attachments to coal, patterns of property ownership, habits of everyday life, representations, and identities. This requires "listening metonymically," being open to the ways that signfiers are articulated but not essentially linked (Ratcliffe 2005, 98; Hall 1988).

Appalachia has often been understood as a resource colony, exploited for the benefit of wealthy outsiders; this is a story with two clearly delineated sides, the oppressors and the oppressed (Gaventa et al. 1990; Lewis et al. 1978). Local supporters of coal are the comprador class. Coalfield politics have been understood in these terms as well—as the famous union song goes, "Which side are you on?" Gaventa's important work on politics in a coalfield community exhibits this dualistic logic (Gaventa 1980). The work's title suggests that there are two potential political dispositions: power or powerlessness. In ways that parallel this book, Gaventa takes on the question of how acquiescence is generated in coalfield communities. Rather than conceptualizing political activism as the aberration, he is instead interested in how it is foreclosed. In one chapter he considers the ways in which "the economic and cultural variables may be interrelated empirically" (131), but his consideration focuses on local "subcultures," identifying certain elements of Appalachian difference as key to the domination of Appalachia. Additionally, the book's definition of politics remains within the public sphere, which focuses the analysis on activities like voting or not voting, participating in organizations and meetings, and the like. *Removing Mountains* extends this investigation into a cultural analysis of space, place, and environmental politics: in particular, how the interdependencies of Appalachia and America relate to the cultural politics of MTR.

The public/private divide structures many readings of political subjectivity. For example, arguing that "fetishisms of locality, place or social grouping" impede the political mobilization of the universal working class (Harvey 1990b, 117) follows the logic of public and private spheres by grouping together the elements of personal life such as race, gender, or place in a secondary position to the primary question of a person's position in the economy, strictly defined (Massey 1994, 134). This economic reductionism often forecloses potential coalitions and leads to political dead ends. For instance, a proposed "Superfund" for workers displaced due to environmental regulation of industry failed because industrial jobs would likely be replaced by service-sector jobs (Gottlieb 1993). Gottlieb suggests that this policy error is related to the different class positions of workers and environmentalists. However, the plan also failed to account for either the geographical or gendered structure of work (Massey 1994). The highly paid jobs in these environmentally destructive industries are the province of mostly white men, and labor market segmentation by gender and race creates a dependent category of jobs in the service sector that provide neither a family wage nor the masculine identity of provider (Hartmann 1976). The intersections of geographic limitations, the unequal labor market, and workers' identities impede a simple transfer to the service sector (Fine et al. 1997; A. Stein 2001). By abstracting away from the territorially embedded and culturally embodied experience of these displaced workers, this proposal posits a false equivalence between market positions.

When a problem is identified as simply economic, it is oddly abstracted in two seemingly contrary ways. "The economy" leads away from the realm of culture, identity, emotion, and meaning at the same time as it conjures an idealized market system, where a system of abstract equivalences stands for and elides material conditions and relationships.[8] In other words, the technical expression of the market in numbers "performs a notion of 'abstract space'" that is a cultural artifact of Euro-American modernity (Thrift 2008, 96; Latour 1993; Lefebvre 1991; Thongchai 1994). The realm of the market is also entangled in a system of gendered dualisms that enhance its cultural significance. Gibson-Graham illustrate the insights that can be gained from expanding the understanding of the "economic" to fields normally considered outside this domain, such as domestic production. Their anti-essentialist class analysis doesn't start from an assumption that only one type of production matters or that the meaning of production is transparent. For example, a male wage worker is working-class in relationship to his employer but may be simultaneously involved in what is more properly termed a feudal class process with his homemaker wife, who herself is implicated in her own class process with the industry that employs her husband (Gibson-Graham 1996). Among other things, this shows how economic relationships carry emotional or affective valences, and how these emotional registers help naturalize systems of domination.

In another example, Steedman recounts how her mother's gendered working-class experiences generated desires for a feminine ideal that eventually led her to support the conservative Tory party (Steedman 1986). Importantly, the seemingly trivial details of everyday life are key to understanding her mother's political subjectivity: the morality of hygiene and feminine style, as well as the dominant middle-class romance of feminine dependency on a male breadwinner. Bourdieu also emphasizes the role of the details of everyday life in creating class dispositions and subjectivity. He describes cultural capital as a set of embodied and materialized class markers; elements of daily life such as furnishings, the number and type of books in the bookcase, and musical taste all contribute to the construction and reproduction of class inequality. The concept of habitus emphasizes the importance of everyday practice; cultural reproduction and political subjectification happen in the ritualized repetition of daily habits including the division of space, bodily comportment, and cultural categories (Bourdieu 1984, 1998).

For some, the deconstruction of the former certitudes of modernity seems to have produced a feeling of hopelessness; they define postmodernity as a confusing pastiche of dislocated signifiers. Jameson's work on postmodernity and cognitive mapping is an example (Jameson 1984; 1991). However, the postmodern schizophrenia that he describes mirrors the fragmented consciousness of the colonized subject, whose reality does not coincide with that of the colonizer (Sandoval 2000, 33; L. Smith 2001, 28). This dissonance opens up the potential for "differential movement" that allows for the "decolonization of meaning through its deconstruction" (Sandoval 2000, 114). Consider the breakdown of the universal and unitary subjects of liberalism and Marxism, representing on the one hand pure willfulness (P. Williams 1991, 219) and on the other the most essential truth of human activity, commodity production (Harvey 1990b, 123). In a classic instance, Gaventa describes a state of affairs in which workers are powerless because their subjectivity is multiple, or fragmented (Gaventa 1980, 19).[9] This fragmentation is a result of oppression; the powerful presumably do not experience multiple consciousness and are able to act politically to pursue their (singular) interests. When workers identify with their bosses, this is called false (or multiple) consciousness—but the bosses are presumed to know what they are doing. However, this seemingly unitary purpose itself begins to crumble when cultural firewalls, especially that between humanity and nature, begin to fail. The notion of a singular capitalist interest—profit—appears more and more nonsensical as the effects of the globalization of modern industry become plain.

Instead of relying on a liberal ideal of a unitary, rational subject, which is modeled after all on people in denial of how much they are products of their so-called dependents, poststructuralist theories of the subject emphasize to what extent all identities are coalitional, partial, fragmented, and located in history (Butler 1990; Sandoval 2000; A. Smith 2008). A subject achieves the appearance of stability through ritualized repetitions disciplined by regulatory fictions. Subjects are not repressed but actually produced by intersecting power structures; the discursive production of subjects thus generates an excess of meaning that refuses simple determination (Butler 1990, 3; Foucault 1976). Political subjects are mobilized through the articulation of numerous and sometimes conflicting identities, such as taxpayer, housewife, and national citizen (Hall 1988). Identity and subjectivity are less fixed characteristics of persons than a set of cultural projects, practices, and formations; this understanding emphasizes the open-endedness of social structures. The surplus of meaning or overdetermination of social things requires a reckoning with ghosts, or an engagement with the emotional, affective, or precognitive traces of things that are marginalized, erased, or exterminated (Berlant 2008; A. Gordon 1997; K. Stewart 1996). This type of theorizing, which Thrift has called nonrepresentational, is therefore cognizant of the role of (representational) social theory in fixing, and thereby denying, the "onflow" of life (Pred in Thrift 2008, 5). For many, the porous contingency of identity formations, which can enable unexpected collaborations, is a source of optimism (Sandoval 2000; Slack 1996; A. Smith 2008; Tsing 2005).

Far from being a rejection of materialism, these theories of subjectivity follow distinctly in the Marxian materialist tradition but expand the range of the political to include those elements of everyday life often naturalized in the private sphere. Joan Scott's definition of discourse reflects this materiality; for her, discourse includes structures, categories, and beliefs. This means that discourse encompasses the institutions normally separated into economic, political, social, and cultural fields and allows an understanding of how they are interrelated and mutually constituted and how they structure conditions of possibility (Scott 1988). Butler uses the term *materialization* to emphasize how cultural categories gain traction by their repetition in everyday embodiment (Butler 1993). Materialization for Butler resembles Bourdieu's habitus: it is a process of stabilization "over time that produces the effect of boundary, fixity and surface we call matter" (9). It is precisely this emphasis on materiality—of the body, the house, and the neighborhood—that leads to a consideration of the cultural significance of geography, the built environment, and economic production.

Hence, my aim is not to take the characteristics of coalfield residents at face value and thereby understand them as coalfield residents per se but instead to understand how they are constructed and construct themselves as coalfield residents and how the discursive structuring of their subjectivity shapes their environmental politics. This requires moving between fields, especially examining how cultural representations and identities are implicated in the construction of the economy, patterns of everyday life, and the human relationship to nature. This is an anthropological notion of culture as a meaning-making system expressed in discourses and bodies, practices and material things that is characteristic of human life. Understood in this way, MTR is a deeply cultural act, and the complex environmental politics of coal mining are, in part, struggles over the meaning of the practice. Although social analyses of mining are usually limited to the economic and political fields, MTR and coal mining in general are enmeshed in networks of materialized signification, including ideas about private property and patterns of property ownership, gender, race and class formations and dispositions, national identity, and the civilizational discourses of modernity. Especially important to this analysis are heteromasculinity and whiteness, which I am considering not as essential traits of coalfield residents but as historically located social projects. Gender and race, through their references to human bodies, signify social structures in a basic, materialized way. These social categories are formed through historical processes and represent a fundamental platform for political conflict and transformation (Hall 1996; Omi and Winant 1994).[10] Subjects are "constituted through the force of exclusion and abjection," which produces for the subject an "abjected outside" constructed from the subject's "founding repudiation." These processes of abjection, "under the sign of the 'unlivable,'" create a "zone of uninhabitability" that "constitutes the defining limit of the subject's domain" (Butler 1993, 3). Dominant, unmarked identities are formed through repudiations and reversals and constantly require the invocation of abjected Others for their maintenance (Kimmel 1994; Messner 2000; Pascoe 2007; Plumwood 1993; Ware 1992). Further, the concept of intersectionality teaches us that these exclusions or repudiations occur at the same time in overlapping and contradictory ways, because identities are multiply constituted via many intersecting categories (Alarçon 1997; Anzaldúa 1987; Hill Collins 2000).

The implications of intersectionality go beyond checking elements off a list (of race, class, gender, etc.) to highlight the importance of cultural analysis (Hartigan 2005). This sort of analysis works against dichotomous reasoning that suggests a person or action can be definitively judged as essentially one thing or another. Instead, social life unfolds in a field of ambivalent meaning and porous categories in which actions can contain productive contradictions and gaps that depend on context, place, and situation (Thrift 2008). In order to understand the ever-changing and yet resilient nature of social hierarchies like race, gender, and class, it is necessary to pay attention to the basic cultural processes of categorization and signification that operate through them (Hartigan 2005; Hill Collins 2000; Omi and Winant 1994). Race, gender, and class are best understood not as properties of individuals but as social projects, or "means of organizing and interpreting everyday life" (Hartigan 2005, 192). In addition, these categories are not only formed in relation to other *human* subject positions and identities but like all forms of human identity are formed in relationship with the environment and natural world (Deloria 1999; LaDuke 2005; Plumwood 1993; Rocheleau et al. 1996; Salleh 1997; Warren 1997).

This question of how the human relationship to nature is implicated in human domination has been addressed through the lens of political ecology and environmental justice (Bullard 1993; D. Moore et al. 2003; Myers 2005) and ecofeminist theory (Adams 2000; Barry 2001; Merchant 1995; Mies and Shiva 1993). These studies have mapped connections between seemingly divergent categories like land use and race, food and gender. But without a serious reckoning with the contingency of the identities in question, this type of study can reify the categories it critiques. Focusing on one element of identity, such as race or gender, to the exclusion of other relevant identity formations can create a deterministic narrative of winners and losers that oversimplifies the situation and often leads to political impasse. For example, a dualistic conception of race that defines things simply as racist or antiracist leads to a reductive definition of whiteness consisting of a series of ahistorical oppositions that erase the complexity and volatility of racialization and its interaction with other social formations (Hartigan 2005, 238). This schema fails to account for the experience of many poor and otherwise marginalized "whites," such as white Appalachians, who often experience privilege and oppression at once, and for the role of place and other cultural forms in determining their experiences.

A cultural analysis of the complex interactions between these objects— race, gender, class, and nature—requires respecting their irreducibility while recognizing their mutually productive effects in the social world. Whiteness, for example, might be productively viewed as a "charismatic traveling package," rather than a definitive state of being (Tsing 2005, 227). As Hartigan notes, whiteness is "hardly a given property of white bodies" (Hartigan 2005, 178). This signals two important insights: race exists in two states at once, the conceptual and the material, and whiteness is a regulatory fiction policed in part through the specter of failed whiteness, or white trash. "White trash" is a trope that disciplines the white subject through a variety of related cultural forms (norms of etiquette and decorum, discourses of degeneracy, and spectacles of deviant sexuality, for example). White trash is a cultural "zone of uninhabitability" that performs the cultural work of reinscribing the category of whiteness (Butler 1990, 3). Although whiteness is ultimately about racial oppression, white hegemony is maintained in part through intraracial struggles on apparently nonracial terrain, as well as through directly racial conflict. Thus, racism is a necessary but insufficient condition for the persistence of race (Hartigan 2005, 257). Stoler draws on Foucault's gestural work on race to suggest that the complex imbrication of race and class is rooted in aristocratic protonationalism in Europe. In the bourgeois state, this discourse of blood, nation, and class was rearticulated in biopolitical form as an internal war defending society against itself and as a colonial strategy of domination (Stoler 1995, 76–79). Because racialization is a cultural process, not a biological one, the continuing social significance of race depends on its pervasive involvement in other cultural forms that constantly echo and reaffirm its logic. In other words, race remains an important cultural tool for domination because it travels and interacts productively with other forms of domination such as gender, class, region, and the human relationship with nonhuman nature.[11]

This book considers the cultural interactions of race, gender, and nature in the Appalachian coalfields, with MTR as an exemplary site. Neither a coalfield nor Appalachia as it is known is a natural characteristic of the territory of the United States. A coalfield is generated by seeking coal, and the production of Appalachia as a region is an ongoing process of marginalization (Massey 1993, 1994; Tsing 1993). The latter is accomplished in part through American national discourses of gender, race, class, and modernity. These processes of subjectification are not only oppressive but also generative of identities and resistance. Race and gender formations intersect with national political culture and regional identity in unpredictable ways that generate the conditions of possibility for environmental destruction as well as environmental justice activism. MTR may resonate with a "white" and masculine relationship to nature, but this relationship is not deterministic. (This is also emphatically not a judgment of individual-level racism or sexism in the coalfields.) Rather, MTR is overdetermined, which is to say that it has an excess of meaning. The practice is enmeshed in contradictory national discourses of hierarchical difference, progress, and citizenship. MTR is therefore able to be an object of contention in various projects around the control of private property, claims to full American citizenship status, and visions of the future.

The Place

The mountains of West Virginia are among the oldest on Earth. Curved by the action of plate tectonics, they are made up of colorful layers of sandstone, shale, limestone, and of course coal. The mountains are also full of water. Fresh cold water bubbles up in springs, often located in small caves that people used to use as natural refrigerators. Icy mountain streams burble over water-smoothed sandstone rocks and pebbles and then disappear into one of the many limestone caves under the mountains. Red and yellow pebbles dissolve into bright pigments when wet, perfect for drawing and leaving messages on the smooth slabs of sandstone that form the beach at Blue Bend National Recreational Area in Greenbrier County. These stones are a witness to the age of the mountains and are dotted with the fossils of ancient plants and animals. The streams are full of crawdads, minnows, and trout; tiny water bugs glide on weightless feet along the surface of the water.

Compared to the Rockies, the Appalachian Mountains are smaller and (perhaps) less spectacular. Nonetheless, there are plenty of breathtaking vistas in West Virginia; from secret places like Buzzard Rock to the New River Gorge Bridge, where thousands visit each year on Bridge Day, when people are allowed to bungee jump off one of the longest steel-arch bridges in the world. Thrill seekers aside, however, what most people remark about the Appalachians is the intimate feel of the mountains. People who grew up there often talk about a feeling of being embraced by the rounded mountains and narrow valleys. The natural beauty of the mountains is often small-scale, too. The forest is mixed hard- and softwood, with maples, oaks, tulip poplars, sumacs, pussy willows, walnuts, hemlocks, spruce, and flowering trees such as dogwood. The forest floor is a damp, loamy mix of leaves and rotted wood where you can find ferns, thistles, and wildflowers, sassafras, ramps (a wild onion), goldenseal, ginseng, and dense thickets of rhododendron. The forest is full of animal life, too, including black bear, white-tailed deer, raccoons, squirrels, skunks, woodchucks, possums, foxes, rattlesnakes and copperheads, bats, frogs, and many, many birds.

A striking thing to a visitor to the state might be the colors. In spring wildflowers bloom, and in summer everything is a different shade of green. Fall brings even more colors as leaves turn brilliant shades of red, yellow, and orange. In winter, the colors are more subtle and varied; the bare gray trees stand out against the black forest floor or against the white snow, and the mountains are dotted with patches of evergreens. Sometimes the rich winter colors are even more beautiful than summer's multitude of greens. A traveler would undoubtedly be struck by the many cutouts on the interstate highways, where the ancient layers of rock that make up the mountains are visible. Wide bands of red, yellow, black, blue, and gray document the epic time and energy it takes the Earth to make a mountain. The top-soil is a rich, moist red. And always, always, there is water, in the under-ground caves, seeping up in the middle of forest trails, in the streams and rivers, and in the rain and fog.

It has often been said that the people of West Virginia are among the nicest people in the world, but the reality is much more prosaic. There are far too many varieties of places and people to make sweeping judgments of that kind. Having said that, one thing a visitor to the southern West Vir-ginia coalfields, in particular, might notice is a generalized preoccupation with saving money that does seem to be idiosyncratic to that region. Store clerks advise shoppers on ways to save money on their grocery bills and then wait for them to exchange the overly pricey items. Food stamp recip-ients draw the judgment of their neighbors when they are seen wasting their allotment on individually wrapped items that are cheaper when bought in bulk. Of course, this is a way of saying that small amounts of money seem to be more important in the coalfields than they are elsewhere.[12] Things that might be disposable in more affluent places are often reused in the coal-fields. Like the so-called hillbilly armor used by U.S. soldiers in Iraq, where undersupplied soldiers have had to create makeshift armor out of what-ever materials they could find (Hirsh et al. 2004), there is a makeshift qual-ity to many of the living places there. Trailers continue to be used long past their life expectancy. In the poorest sections of the coalfields, crumbling roofs are covered with tarps and cardboard.

The natural beauty of the Appalachian Mountains is not cordoned off around the area known as the coalfields, yet that area has a marked lack of public parks and other nature-based attractions, especially compared to neighboring counties. When describing the coalfields, "natural beauty" is not a phrase that comes immediately to mind. A visitor entering the coal-fields on Corridor G, the four-lane highway that goes south from Charleston through Boone and Logan counties, through Williamson in Mingo County, and on into Kentucky, might notice the focus on industrial land use. In 2004 the conveyor belt and dragline being used at the enormous Hobet moun-taintop mine outside Madison were visible from Corridor G, especially at night, when they were lit up like garish yellow Christmas decorations for the twenty-four-hour operation. The periodic flooding that the coalfields have experienced, and which seems to have become more frequent in re-cent years, often leaves debris on the banks of rivers and trash hanging from trees above the streams. Because of a lack of public services like garbage collection, illegal dumping and burning of trash are commonplace. If the services exist, they are prohibitively expensive. Houses in unincorporated areas frequently lack septic systems. Trailers especially are damaged by floods, which doesn't necessarily imply that they are replaced or abandoned. People talk about boom and bust cycles in industries like coal, but the history of West Virginia coal mining is more like a single boom and an increasing bust—at least for the people who live in the coalfields. Coal was discovered in Boone County in the eighteenth century, but it really didn't have an impact on local life until the next century, when the railroads brought capitalists to the area and around 75 percent of the land was acquired by absentee corporate landowners (Eller 1982; Appalachian Land Ownership Task Force 1983). Coal mining required more labor in the earliest years than the small mountain communities could provide, and so the coalfields became a destination for southern and eastern European immigrants in the early twentieth century. Coal mining also offered African Americans a way out of the South, and the early coal mining communities were among the most diverse in America at that time, albeit segregated both residentially and occupationally. Because of this diversity, early unionization efforts by the United Mine Workers of America (UMWA) were racially inclusive, although still characterized by racism and preferential treatment for whites (H. Hill 1988). Coal mining was marked by violent labor struggles until unions were legalized in the 1930s (Batteau 1990; Meador 1981; Savage 1990; Shogan 2004). From the 1930s through the 1950s, innovations in mechanization made mining easier and improved production but still required a large workforce. Demand was high and led to a boom in coal culture. Surface mining began in the 1950s, and since that time the number of miners has decreased by about 80 percent (Caudill 1962; West Virginia Coal Association 2008). The population plummeted as unemployed miners and their families left the area for destinations like Detroit, Ohio, or North Carolina.[13] At the same time, increasing mechanization both on the surface and underground has improved productivity to the degree that, in 2005, a workforce of less than eighteen thousand was able to mine over 17 million more tons of coal than one hundred thousand workers did in 1952. The decades between 1960 and today have seen many small boomlets and busts in mining employment but have been marked most obviously by an increasing divergence between production levels and employment levels (West Virginia Coal Association 2008). In the meantime, many black former miners and their families have left the area, and many of the affirmative action–era trailblazing women in mining have been pushed out or retired, leaving a greatly diminished but predominantly white male workforce in place (Maggard 1999; U.S. Department of Labor 2009).

The Book

As sometimes happens, this research is very close to my heart. My own homeplace is a few miles from the washout, or endpoint, of the Beckley coal seam, a geological accident that made eastern Greenbrier County, where I grew up, much luckier than the west end of the county and even more so than the coalfield counties farther to the west. This statement would appear to be something like a claim of insider authenticity, but my relationship to this research is more complicated than first appears. My insider/outsider status begins with the fact that coal mining was an ever-present but alien world to me growing up in eastern Greenbrier County. The nightly litany of mines on strike was a taken-for-granted backdrop of local TV news in the early 1980s, and in high school, I'm ashamed to admit, I thoughtlessly joined my classmates in mocking the kids from the coal-mining town of Charmco in the west end of the county. I'm an outsider in the sense that almost as much of my life has been spent outside of West Virginia as in it, and an insider in that my research has been a pleasurable process of coming home. As I found in the course of this research, "Where are you from?" is not always an easy question (Visweswaran 1994, 115).

I became aware of MTR in the mid-1990s when I started noticing a certain type of letter to the editor in the *Charleston Gazette,* a liberal-leaning state newspaper often critical of the coal industry. These letters from the coalfields chastised the newspaper and environmentalists for their failure to appreciate the good that the coal industry had done for the state and for valuing nature over people's need to make a living. I struggled to understand the perspective of these coalfield residents. To move the question beyond the obvious reference to jobs versus the environment, I decided to examine what claims each side was making about the coal industry and MTR. I began this research in the summer of 2000, at the height of the hysteria in reaction to Judge Hayden's court ruling that called for the enforcement of the Clean Water Act. At the time, some environmental activists told me that they were not against MTR per se but only against MTR "as it is currently practiced." In the years between 2000 and 2004, the Clean Water Act was rewritten to allow the dumping of waste into streams.[14] When

I returned to the coalfields in 2004 for more fieldwork, it seemed to have become clear to everyone that MTR could not be made to conform to the preexisting environmental laws. MTR is one of those practices occurring at the edge of law that allow for capital accumulation (Tsing 2005). Because it exists on this edge, debate can remain focused on the legality or illegality of particular cases instead of the more deeply political questions of what the practice is doing to the coalfields. The terms of debate may be shifting, however. During my latest trip, in 2008, questions of sustainable energy, global warming, and fossil fuels were in everyone's minds, perhaps because of the presidential campaign.

Like the rest of the United States, the coalfields are torn up with these profound philosophical divides. Some people are convinced that West Virginia's fate is firmly tied to the coal industry's survival, or more specifically to its increasing success. Others see the industry's profit as West Virginia's loss. Another aspect of my insider/outsider status is relevant here. Resource extraction implies connections between margin and center; the inherently dominating relationship between city and country reflects this interdependence (R. Williams 1973). Therefore this book aims "to disclose not an 'other' America, but the one right before our eyes" (Di Leonardo 1998, 24). While the book is deeply concerned with the fate of a particular part of the Earth, the ethnography here is not firmly located in a geographical site. MTR is a cultural practice wrapped up in national and global histories; it cannot be adequately understood only from a local perspective (Kondo 1997; Massey 1993). Therefore, the analysis moves from national representations of the place to local inhabitations. This shifting perspective reflects my own traveling back and forth between West Virginia, California, and eventually Missouri over the course of eight years of research, a journey that has provided opportunities for many fruitful encounters. Over the course of three research trips to the coalfields, I conducted semistructured interviews with environmental activists, miners, ex-miners and family members, county officials, coal-related-industry employees, and other coalfield residents.[15] These interviews were relatively informal and conversational and loosely focused on the coal industry in southern West Virginia, the increasing use of surface mining, and related questions. I found participants through local networks and sought a range of perspectives in order to understand how ordinary coalfield residents felt about the coal industry and MTR.[16] My initial trip, in the summer of 2000, focused on the town of Blair. The majority of the interviews happened in 2004, when

I was able to stay for six months. I also spent time informally with coalfield residents, attended a variety of community events, and visited coal-related sites. In 2008, I solicited participants with a newspaper advertisement that allowed me to reach additional working miners and their spouses. The coal industry and MTR are such controversial issues in the coalfields that I found many people unwilling to participate in my research. Even asking about MTR is difficult; it is normally excluded from polite conversation. There is also a healthy distrust of strangers doing research, due to a longstanding tradition of outsiders portraying Appalachian communities in lurid ways in the national media. Additionally, anxiety over job losses and blacklisting contribute to an atmosphere in which it is safer not to say anything; simply asking about the coal industry is enough to imply a critique. To overcome these obstacles, I stressed my interest in all perspectives and did not push people to answer uncomfortable questions. Nonetheless, my questions elicited strong emotional responses in many cases. In 2004 I found that while the sequelae from the lawsuit in Blair had made it easier for some people to talk to me, it had become impossible for others. In 2008, my participants were all self-selected, and therefore talkative, outgoing people were probably overrepresented.

Rather than revealing "classic norms" that can be used to codify the coalfield response to MTR (Rosaldo 1993, 58–59), these interviews represent some examples of what it is possible to say in a specific place and time; they are expressions of the structures of feeling that animate mining and coalfield activism. As someone raised in West Virginia, my semi-insider status gives me a seemingly privileged view into the truth of people's perspectives; being from Greenbrier County did help me gain some people's trust. However, we can "never wholly understand and identify" with others on the basis of a constructed, coalitional identity (Visweswaran 1994, 100). Indeed, my supposedly insider status is belied by my very traveling back and forth—a journeying that uncomfortably replicates the epistemological positioning of colonial anthropology (102–3). This failure is inherent in representations of social worlds, yet acknowledging this uncertainty "takes us homeward not away" (99).

My initial research was organized around the town of Blair, the site of the controversial Daltex mine. I expected there to be two sides to the issue, an old Appalachian side of people who wanted to stay in their homeplaces and a workers' side of people economically dependent on the industry. What I found was less clear-cut. In fact, the more I talked to people, the more questions I had. How can people simultaneously love a place and support its destruction? How can two groups of people see the same landscape in entirely different ways, as a wasteland or an Eden, as an abomination or a site of technological improvement? How do people reconcile MTR with the famous Appalachian attachment to place? The politics of MTR are tied up in struggles over the meaning of the place, the meaning of coal mining, and around people's fears and hopes about the future. These are politics in which a person can hold a critical view of industry "propaganda," be an environmentalist, support workers' rights, and simultaneously treat the coal industry as the water she swims in. These are politics in which the core beliefs of American political culture—the sanctity of private property, independence, and progress—both require and produce a region's poverty, dependency, and destruction.

Because this book is framed by the production of Appalachia as an American region, it vacillates between representation and inhabitation, the micropolitics of daily life and longer-term trajectories. In these chapters there is a recurring concern with temporality, both in terms of Appalachia's place in an American progress narrative and in terms of West Virginia's history as a social body. There is also a sense of urgency in the writing of this book. MTR is a moving target, swiftly remaking both the literal and figurative terrain in which coal mining happens. Therefore, the text might be read as a montage of traces left by a history of resource extraction (K. Stewart 2007; Buse et al. 2005). The past (and the past-to-be) haunts the present here in the form of a nostalgia for something that may or may not have existed but which is rendered more and more important by its impending erasure by the increasingly rapid extraction of coal. The ephemeral mountain and coal cultures slip through the fingers of both activists and industry boosters, as they struggle to hold on to a position from which to speak. The object of this book is the coproduction of meaning and economy in the practice of MTR. Instead of providing a final answer to the question of what MTR means, or how coalfield residents feel about it, the book offers impressions of some of the key cultural processes at work in and around the practice.

I set the stage by examining the production of Appalachian identity in a variety of national cultural sites in chapter 1, "Hillbillies and Coal Miners: Representations of a National Sacrifice Zone." Bringing together divergent registers of representations illustrates the repetition of certain tropes of Appalachian difference. Appalachia is imagined in racialized and gendered terms that help construct and reproduce a particularly American narrative of differential development and sacrifice. Focusing on the local meanings of mining, chapter 2, "Men Moving Mountains: Coal Mining Masculinities and Mountaintop Removal," examines the magic of labor subjectification in capitalism, which is capable of convincing people to do improbable things and is visible here in a series of gender stories coalfield residents tell that inscribe mining into culturally legible contexts. The transformations that the increasing mechanization of mining is bringing about are also transforming the ways that coal mining makes sense, as the stories shift from intimate ones of relationships and physicality to a more modern technological mastery of nature.

Reconsidering these local meanings through a wider lens, chapter 3, "The Gendered Politics of Pro–Mountaintop Removal Discourse," examines the role of American political culture in constructing and gendering notions of independence and dependency. White male working-class subjects must negotiate a space of freedom between two threats to their status of independent citizenship: a state of natural, feminized dependency on one hand and the oppressive interference of the state in their natural freedom on the other. These gendered concepts give jobs in mining more than strictly economic significance in the coalfields.

Because the book is concerned with local meanings and inhabitations, it is interesting to note that the all-terrain vehicle, or ATV, turns out to be central to struggles over land use and coalfield identity. ATVs allow coalfield residents access to the undeveloped corporate-owned land that they have effectively treated as commons for most of the last century, where they have been able to hunt, fish, and gather forest products. Recently, a public-private partnership attempting to rationalize and commodify ATV use in the coalfields has disrupted local land use patterns. Chapter 4, "ATVs in Action: Transgression, Property Rights, and Tourism on the Hatfield–McCoy Trail," exposes the divergent local notions of property rights that appear in conflicting narratives of Appalachian intimacy with the land and dreams of development and Americanization.

Dreams of the future are necessarily predicated on interpretations of the past, and chapter 5, "Coal Heritage/Coal History: Appalachia, America, and Mountaintop Removal," analyzes two competing versions of the coalfield past, one represented by the coal heritage movement and the other represented by preservationists and environmentalists. Coal mining can be told as an exemplar of an American story of progress and technological development or as a story of social injustice and conflict. MTR is seen as the latest iteration of either progress or hyperexploitation. The chapter maps the linkages between each of these visions of history and the divergent futures imagined and hoped for by coalfield activists and industry boosters. Ideals of economic development, class identity, and gendered heteronormativity each make sense within a national culture that is deeply shaped by histories of race-based exploitation and racial distinctions. Chapter 6, "Traces of History: 'White' People, Black Coal," examines the recurrent themes of temporality, ideas of progress, and American cultural citizenship through the lens of American racial formations. Whiteness functions as a regulatory fiction in struggles against Appalachian cultural marginalization. Appalachian Otherness renders coalfield whiteness problematic. Those white coalfield residents desirous of unmarked American identity are required to negotiate their identities through discourses of differential worthiness, especially articulated around hygiene, standards of living, and work. The cultural content of this racial project structures the societal relationship to land and limits the terms of environmental and political discourse.

The Conclusion uses gendered and racialized imagery from the 2006 *Coal Facts* brochure published by the West Virginia Coal Association to bring together the themes of the book and consider alternative possibilities. Coal industry supporters imagine the land in reductive ways that support a hierarchical and dominating dualistic view of humanity and nature. In this context, anti-MTR activists use metaphors of indigeneity to articulate a different way of relating to the land. While in some ways replicating the colonizing logic of dominant American culture, these metaphors also indicate the potential for powerful new coalitions around environmental justice.

Making sense of MTR requires moving beyond a strictly political-economic account to a consideration of the "intimate public sphere" (Berlant 1997). That is one reason this story depends on exploring unexpected connections and disjunctures. The figure of the coal miner relates to national imaginaries of work, abjection, and progress, as well as to notions of gender and family relations. Similarly, ATVs connect narratives of rural backwardness and frontier masculinity to the daily experience of inhabiting a place and a modern, technologically mediated experience of nature. MTR makes sense in the context of certain notions of time, ideas about differential progress and development, and teleological metanarratives of national identity. These temporal narratives gain traction in part through their gendered and racialized content. For example, overtly rational corporate

View of reclaimed flattop, Blair, West Virginia, 2004. Photograph by the author.

practices draw on liberal notions of formal equality but are based on substantive inequalities of property rights and legal protection that recall the
centrality of whiteness in American culture and national identity (Hartman 1999; Lipsitz 1998; Wiegman 2003).

Appalachian cultural and economic marginalization operates as part of
an exclusionary and hierarchical system of civic morality and cultural citizenship and produces unexpected political subjectivities and relationships
to nature. MTR makes sense, for some coalfield residents, as the optimal
use of the land, according to the national/global so-called free market. The
flattened and deforested landscape is the word, of the corporate profit imperative, made flesh. It is an expression of "hyper-rationality" that sees unitary liberal individualism as the ultimate signifier of civilization (Cuomo
2005, 199). This individualism allows a "lofty and distant contemplation"
of the transformed Appalachian landscape (Myers 2005, 43) that is objectified as property and reflects that formative repudiation that casts nature
as Other to the reasoning modern self.

CHAPTER 1

Hillbillies and Coal Miners:
Representations of a National Sacrifice Zone

"A Strange Land and Peculiar People"

Thomas Jefferson's *Notes on the State of Virginia* contained a full accounting of the natural resources available in the part of the world that would become West Virginia, including the coal (Jefferson 1794; Myers 2005). His catalog of the riches awaiting the colonizers of this territory foreshadowed the way that the region would become a sacrifice zone for the nation; the coal in the Appalachian Mountains is America's coal reserve. Today the national interest is mobilized to support the coal industry against its critics; in a time of energy crisis and war, the argument goes, it is time to put aside differences in the interest of our national security. This is where the contradictions of national interest become apparent. Appalachians are constantly asked to sacrifice for the nation to which they belong, but this belonging is problematic. As prototypical white rural citizens, they are in some senses ideal Americans, but at the same time they are culturally and economically marginalized, and the national/corporate interests they are asked to serve are not necessarily compatible with the survival of their communities and practices.

A sacrifice zone is a place that is written off for environmental destruction in the name of a higher purpose, such as the national interest (Kuletz 1998). Although by their very nature environmental hazards defy social and cultural boundaries, environmental exploitation relies on the maintenance of such imaginary firewalls. A system that depends, contradictorily, on the destruction of its own natural basis (O'Connor 1994) can only survive because those who benefit feel safe from the type of environmental and economic disaster experienced by the poor.

This chapter offers a guided tour of the cultural production of Appalachian identity, which contributes to the region's construction as a sacrifice zone. Stories about the region constitute a genre, in the sense of a set of conventional narratives relying on certain stock characters and affects that can be "repeated, detailed and stretched," containing productive con-tradictions (Berlant 2008, 4). This genre crosses widely divergent regis-ters of cultural production, from the epic to the trivial, including scholarly writing, professional journalism, television, cinema, and B-grade horror films. To demonstrate the flexibility and pervasiveness of this performative genre, I focus on representations of (in)famous West Virginians Jessica Lynch and Lynndie England; the films *Deliverance* and *Wrong Turn;* a book of photojournalism, *Coal Hollow;* cable television news coverage of the region; a situation comedy; and a *Saturday Night Live* skit. Rapidly shift-ing from one register and type of representation to another denaturalizes the web of buried connections that gives the region meaning. Despite their differences, each of these representations of Appalachia becomes legible through an iteration of a set of ideas about the identity of Appalachia as a region that reaffirm its fundamental difference and its suitability for sacri-fice. Stereotypes instruct about who belongs and why. They are a system of ordering the world that is able to convey a large amount of information simply and rapidly about dominant social values (Dyer 2000). Like other acts of signification, representations of Appalachia gain meaning through their relationship to other signs in a network of articulations that reaffirm and support each other (Hall 2000). These representations are part of a web of citationality that reproduces the Appalachian region as a conven-tional part of the geo-body of the nation (Thongchai 1994). By ceaselessly connecting poverty and environmental destruction with fundamental dif-ference, these representations reaffirm the imagined immunity of the main-stream from this type of suffering.

Since the nineteenth century, travel writers, novelists, social reformers, and humorists have represented the area for a national audience, portray-ing the mountains of central and southern Appalachia as a wild, inaccessi-ble wilderness, a region of people left behind by modernity, distant in time and space from the everyday world of American society (Billings et al. 1999; McNeil 1995). In the eighteenth to mid-nineteenth centuries, remote mountain communities offered a space of racial ambiguity and freedom. The Melungeons were an indeterminate group believed to be related in varying amounts to Native Americans, African Americans, and Portuguese. Increasingly stringent race-based laws probably led to their absorption into strictly enforced legal categories that gave some of them rights to their land and expropriated others (D. Wilson and Beaver 1999). By the late nineteenth and early twentieth centuries, accounts of Appalachia refer to

Appalachians as "our contemporary ancestors" and as "pure Anglo-Saxon stock" and the region as "Elizabethan America" (McNeil 1995, 91, 150, 206). These mythologizing accounts of Appalachia describe people, material culture, and dialects as though descended virtually "unadulterated" from Britain. Significantly, the representation of Appalachians as primitive Anglo-Americans also performs a claim to territory based on indigeneity or ancestral ties that mimics Native American claims.[1]

The pattern of representing Appalachians as white reveals some interesting things about the transformation and persistence of race and class inequality in the United States.[2] Recurrent stories of white trash, hillbillies, and other poor whites reflect the dissonance between ideal images of American citizens and the reality of lived inequalities. Poor whites are matter out of place in the cultural map of the United States, which identifies poverty with blackness and wealth with whiteness (Douglas 2002; Hartigan 2005; Wray 2006). Representations of these poor whites have historically relied on quasi-racializing terms, in which physical characteristics such as skin or hair are seen as especially telling (Hartigan 2005, 72). Perhaps "white trash" performs similar work to the "specter of the fag"; thus, whiteness would be understood as a fragile identity, put at risk by lapses of etiquette or decorum (Hartigan 2005; Pascoe 2007). Like the fag discourse, which instructs boys on acceptable masculinity, white trash discourse does more than just stigmatize poor whites. It also disciplines an ideal "white" subject through these processes of repudiation and abjection. These "disidentifications" reaffirm the logic of differential worthiness, which teaches that those who cannot make it are themselves to blame, and by extension, the wealthy deserve their comfort and are safe from the fate of the poor (Ratcliffe 2005, 62).

But hillbillies are not necessarily white trash. Unlike white trash, hillbilly pride is possible. The latter term connotes not simply embodied filth and abjection but an array of cultural forms related to the past (Hartigan 2005, 124). Indeed, representations of the Appalachian region also frequently suggest a *place* forgotten by time. The case of Appalachian cultural marginalization is particularly interesting from an environmental perspective because of the axiomatic identification of the people with the landscape. Appalachian studies scholars have worked hard to combat persistent negative stereotypes about Appalachians (Billings and Blee 2000; Billings et al. 1999), but there is an equally long tradition of glorifying Appalachia as a bucolic landscape and white Appalachians as ideal rural citizens (Batteau

1990). Whether imagined in valorized or debased form, the heterogeneity of Appalachian life is erased in the processes of Appalachian marginalization. "Appalachia" is embodied in particular racialized and gendered figures: the hardworking coal miner, the indolent hillbilly, the "rural sodomite" (Bell 1997), and the oppressed mountain woman.[3] At the same time, Appalachian virtue and backwardness are often articulated with representations of Native Americans in the national culture.[4] These complex representations of morality and difference contribute to the construction of a resource colony—a persistent frontier in the nation's heartland.

Liberal democratic citizenship is characterized by a contradiction: citizenship is ideally based on the universal equality of humankind but is contingent on national histories characterized by hierarchy and domination (Kazanjian 2003). American citizenship is shaped by a set of ideals articulated with whiteness in U.S. racial formations: property, progress, centrality, and independence.[5] The modern man's self-conception "as free, productive, acquisitive, and literate" puts in place a logical framework for judging the acceptability of racialized groups for citizenship in a modern state (Goldberg 2002, 109). According to these racial logics, whiteness "in class terms . . . definitionally signifies social superiority, politically equates with control, and economically equals property and privilege" (113). These racial logics contribute to "the colonizing trick," or the ranking of human populations in these terms (Walker 1829; Kazanjian 2003). White trash, hillbillies, and rednecks serve as a kind of battleground for American racial politics, representing a space where racial order is both asserted and challenged. Appalachians are symbolically coded white in U.S. racial formations, but this marked whiteness is problematic. Appalachia is seen as a place apart; disidentification hides the essential role of the place in the national and global system (Fisher 1990). The dynamics of these representations help create a national sacrifice zone, a place where the sacrifice is at the same time compelled and freely given.

Jessica and Lynndie

Narratives of Appalachia focus on two stock character types, either pure white, American stock—the best of America and for America, a kind of natural human resource for the nation, an ideal white Christian rural citizen; or a degenerated white, rendered impure either through inbreeding, poor ancestry, or simply through geographical isolation. Jessica Lynch and

Lynndie England, two West Virginians whose lives have been forever linked to the American war in Iraq, illustrate the two sides of Appalachian whiteness in these national narratives. Their representations in the media express the idealization and demonization of Appalachians in American culture and demonstrate the work these representations do in making claims about the American national character.

Jessica Lynch was the most famous prisoner of war of the war in Iraq. In November 2003, she was on the cover of *Time* (Gibbs 2003). She became the blond, wholesome face of the U.S. invasion of Iraq. Her (white) innocence—even after the story broke that the Pentagon had sensationalized her "rescue" for propaganda purposes—was counterpoised in the media with the guilt of her (dark) Iraqi captors, who, it was claimed, had treated her brutally, though she had no memory of it (Colford and Siemaszko 2003). The *Time* article offered a glimpse into the bucolic hometown life that had produced someone like Jessica Lynch, "just a country girl" whose dreams in life were to join the Army, become a kindergarten teacher, and start a family. In this story, Jessica's hometown county fair represents a relic of America's idealized past, safe and secure, neighborly, with no crime, drugs, or urban corruption. This imagining of rural West Virginia as America's past was replicated in the made-for-TV movie *Saving Jessica Lynch*, which aired on NBC in 2003 and shows nothing of West Virginia except in the homecoming scene, in which Jessica receives a hero's welcome in a parade down a hazily figured American Main Street lined with red, white, and blue bunting.[6]

Media accounts of Jessica's life did not portray her West Virginia origins as exceptional. In these accounts West Virginia is merely rural America, no different from Iowa or any other rural state. Jessica seemed to represent not West Virginia or Appalachia but the good-hearted, simple American youth who were fighting the war in Iraq. This role would not be remarkable, except that another prisoner of war from Jessica's unit, Shoshanna Johnson, who is black, received notably less attention from the national media. Jessica's whiteness was part of what made her story work; the Jessica Lynch story, as told by the Pentagon and obliging media outlets, produced her as the "White Woman" who needs rescuing from the "Man of Color" (Frankenberg 1997). For Jessica Lynch, Appalachia was not a stigmatizing element of her identity—rather, it was simply one element of her media image of an ideal American rural youth, sacrificing for the nation. It is precisely as part of this idealized white citizenry that Appalachians can escape their marginalization in the national media.

Lynndie England, however, caught smiling in photographs documenting the human rights abuses at the U.S. military prison Abu Ghraib, ended
up representing the dark side of rural white America, especially Appalachia,
in media representations of her life. Unlike Jessica, whose decision to enlist
in the military to pay for college was interpreted as a sign of her independent character, Lynndie's actions were negatively interpreted, as in early
reports on Abu Ghraib in which her decision to enlist was portrayed as an
indication of simplemindedness.

Media accounts of Lynndie's life frequently referred to her "trailer" background, as if this added important information to the story of Abu Ghraib.
Reporters repeatedly remarked that Lynndie, her family, and people in
Lynndie's community live in trailers, often mentioning this fact in incon-
gruous places, as in a story from the Scottish newspaper the *Herald,* which
noted: "In the previous photograph, the 21-year-old *from a trailer home in
West Virginia* was seen, cigarette in mouth, giving a thumbs-up sign and
pointing at a humiliated, naked, hooded Iraqi prisoner" (Simpson 2004,
my emphasis). Lynndie's hometown was accused of harboring the Klan,
in an apparent attempt to explain her actions (Caldwell 2004). Even more
sober assessments of the events at Abu Ghraib seemed compelled to bring
the Appalachian element into their discussion of Lynndie's crime. A reporter
for *Newsweek* managed to simultaneously evoke and dismiss the cultural
imagery of the hillbilly: "Some have wondered whether England, who
came from a poor town in the hollows of West Virginia and lived for a time
in a trailer, was the victim of a deprived or degraded socio-economic background. But her parents appear to have been loving and her childhood innocent" (Thomas 2004). Although Lynndie's childhood appeared innocent
to the reporter, her geographical origins did not. As a scapegoat for U.S.
crimes in Iraq, Lynndie took the fall for the failures of American whiteness; she became the embodiment of exactly what the war in Iraq was not
supposed to be—racist, brutal, and ignorant—and displaced that guilt from
the war's initiators. In this way, Lynndie is simply a recent example of the
way that the figure of the ignorant redneck is a displacement of white
American racial guilt (Graham 2001). But Lynndie's image in the media
also indexed a whole range of specifically Appalachian stereotypes—mental deficiency, oppressed femininity, irrationality, and perhaps even a hint
of the monstrous (Kristeva 1982).

Jessica and Lynndie illustrate the two faces of Appalachian whiteness
in American national culture. On the one hand, (white) Appalachians are

imagined as ideal rural citizens, independent, brave, honest, and most important, always ready to sacrifice for the good of the nation. These idealized Appalachians are understood to be good-hearted, fair-dealing people, usually racially innocent (innocent of "other" races), all qualities stemming from their isolation from modernity (Inscoe 1999). They have pure, country lifestyles and are frequently constructed as bastions of Christian piety in an otherwise secular society; these would be the "real Americans" from the 2008 Republican presidential campaign. On the other hand, like other poor southern whites, Appalachians are demonized through the usual stereotypes of stupidity, backwardness, racism, sexism, and uncontrolled violence. In the case of the hillbilly, however, the mountainous landscape itself is implicated in the degeneration of the people who inhabit it. Appalachian stereotypes conflate the land and people, with dark, trashfilled hollows sheltering isolated, incestuous communities. These generic representations are instrumental in the cultural, social, economic, and environmental marginalization of Appalachia.

Appalachian Poverty: Backward Hillbillies or Exploited Coal Miners?

The two most iconic images of Appalachians in national culture are probably the hillbilly and the coal miner. In the hazy imagination of the region in popular culture, the two are often confounded, despite their seeming disparity. A coal miner is after all an industrial worker, a very modern thing (though dated in a postindustrial economy). And a hillbilly is the very embodiment of backwardness and isolated rurality. Perhaps because of their physical closeness to nature, the two images easily converge, as in the 1980 film based on Loretta Lynn's life, *Coal Miner's Daughter,* which drew on both tropes: Lynn's large, impoverished family, living in a cabin up a mountain "holler"; her hardworking coal miner father; her underage, semi-abusive marriage; her life and her mother's life as excessively fertile and oppressed women; etc.[7] Loretta Lynn's stage persona also reflects the confluence of hillbilly and coal miner. At performances, she wears frilly old-fashioned gowns with high necks and long sleeves that evoke the image of a deeply religious mountain femininity from the preindustrial era. This suggestion of traditional and antifeminist femininity complements the confluence of the coal miner and backward hillbilly masculinities. At the same time as they contribute to the production of marginalized Appalachian

identity, these iconic symbols of place recenter the masculine national subject and affirm the auxiliary status of women, as Loretta Lynn's title indicates.
One thing the hillbilly and the coal miner have in common is their relative poverty, and given the association in American racial formations between whiteness and wealth, this is one reason they are problematic whites.
Pre-coal, people in the Appalachian region were not all that poor in absolute
terms. Appalachia had both a self-sufficient subsistence farming economy
and many small industries based on salt and forest products (Billings and
Blee 2000; Precourt 1983; Pudup 1990; Pudup et al. 1995). However, the
noncommodity consumption patterns of subsistence farming are read as
signs of misery and abjection in American national culture. Well water,
outhouses, homespun cloth, newspaper-covered cabin walls, and other
"old-timey" artifacts of Appalachian material culture bear the stigma of a
lack of integration into the market economy. Nonmarket consumption
patterns signify a lapse in modernity; they reaffirm the region as a "place
on the side of the road," an exception to mainstream American culture
(K. Stewart 1996).

A recurring skit on NBC's *Saturday Night Live* depicts this kind of
stigma. Like many skits on the comedy show, "Appalachian Emergency
Room" is a parody of a TV show. The introductory image is washed out,
like an overexposed photo. It is a scene of a mountain landscape (from the
western United States, with a yellowish hill dotted with scrubby trees, hazy
mountains in the distance). Regardless of geographical inaccuracies, the
image places the skit and evokes the familiar identification of the Appalachian landscape and people. Appalachian difference is further signified
as the scene shifts from the mountain landscape to an ambulance traveling
through the woods, with the name of the show spelled out in rough wooden
letters while a banjo plays.

One episode of this skit opens with a triage nurse (regional identity
signified by his accent and his mullet) calling the next patient, "Percy Bo-
dan."[8] A couple shuffles to the desk, the man in overalls, with a bag of
frozen peas duct-taped to his head, the woman wearing a housedress, slip-
pers, and socks. Percy explains what happened to his head, "Since it was
pretty out, we was sitting on the finished side of the porch . . . with our feet
in the bathtub." As the live audience laughs, the woman adds, "We got an
outside bathtub," in a voice meant to indicate mental deficiency. Percy
was injured when he fell into a pile of old radiators, where a raccoon bit
him on the head. The woman mentions her son's "Filipino wife," who "said

something real hateful in Filipino" and how the raccoon that bit the man's head "got all riled up."

Other patients repeat the pattern of stupidity, one getting stuck inside the glass case of a claw-toy-grabbing game, another getting an action figure stuck up his rear end at a yard sale, another injuring himself sliding down a broken pool slide where there was no pool. Another patient is simply alcoholic. All are dressed in outdated clothes, including tight cutoffs, old Indy 500 T-shirts, and other styles meant to suggest the passé. The exchanges are united in mocking the laughable amusements of the poor. "Sitting on the finished side of the porch," playing on broken pool slides, or gathering "a box of old Barbie doll heads" for a yard sale are all examples of a pathetic approximation of fun, often derived from trash. The skit draws a familiar equivalence between poverty and stupidity, especially with a southern accent. Percy Bo-dan's case highlights the ambivalence between "native ingenuity" and the pathos of poverty. They arrived at the hospital, the woman says, riding on their lawn mower. The skit manages to hit on a wide variety of other Appalachian stereotypes as well, including white racism and xenophobia, alcoholism, and sexual deviance. But the most striking theme from the skit is the stigmatized consumption patterns and material culture of the poor.

Appalachian difference had been a popular topic for travel writers and others since the mid-1800s, but it was not till the onset of industrialization, from about 1890 to 1930, that images of Appalachian poverty received national scrutiny. At this point, the Appalachian subsistence economy was depicted in a dichotomized way; infrequently, it evoked the ideal self-sufficient white American pioneer from frontier days, but more often it was assumed that the Appalachian's distance from the market signified stagnation and backwardness (Precourt 1983, 99). Strikingly, the living conditions of white Appalachians were seen as comparable to those of African Americans living in slavery (Klotter 1980). Nonmarket consumption patterns were especially notable. As travel writer and social reformer Ellen Churchill Semple remarked in 1901, "Sugar is never seen in this district, but backwoods substitutes for it abound ... [including] honey, sugar maples, sorghum molasses" (1901, 155). Thus the original sweeteners of the temperate zones, honey and maple syrup, are regarded as substitutes for the global commodity (Mintz 1986).

This stigma and scrutiny given to Appalachian culture was shaped by U.S. racial formations (Omi and Winant 1994). Appalachian poverty, whether absolute or relative, has consistently been constructed in American culture as exceptional white poverty (Klotter 1980). Semple emphasized that the "Appalachian" was a particular kind of Other: "Their manners are gentle, gracious, and unembarrassed, so that in talking with them, one forgets their bare feet, ragged clothes, and crass ignorance, and in his heart bows anew to the inextinguishable excellence of the Anglo-Saxon race" (1901, 151). Despite this "excellence," Appalachians were at the mercy of the landscape. Again and again, the geography of the region is identified as the root of the problem. As Semple phrased it, "In consequence of his remoteness from a market, the industries of the mountaineer are limited. Nature holds him in a vice here" (159).

Then came coal. Once again, nature gave hillbillies the short end of the stick, as Berea College president William Goodell Frost put it: "The jackals of civilization have already abused the confidence of many a cabin home. The lumber, coal and mineral wealth of the mountains is to be possessed, and the unprincipled vanguard of commercialism can easily debauch a simple people. *The question is whether the mountain people can be enlightened and guided so that they can have a part in the development of their own country or whether they must give place to foreigners and melt away like so many Indians"* (1899, 105, my emphasis). Suddenly exposed to the roughest face of the economy that they had shunned, backward hillbillies became downtrodden coal miners, leading Kai Erikson to conclude that they were "among the most truly exploited people on Earth" (1976, 68).[9]

In *Night Comes to the Cumberlands,* probably one of the most influential books of the twentieth century on Appalachian poverty and environmental problems, Harry Caudill paints a grim picture of the Kentucky coalfields in the wake of mechanization and strip mining. The book presents a strong critique of the coal industry's hyperexploitation of the area, yet it tempers this critique with an assumption of basic differences that are seen to exacerbate the economic crimes of the industry. Beginning with an overview of the British criminal element that he claims settled the hills, Caudill's portrait reiterates many traditional representations of "our contemporary ancestors" (1962, 77). Children "reached maturity with the impetuosity of Indian braves," and "hospitality was the mountaineer's noblest virtue" (78). At the same time, "Sexually, the mountaineer enjoyed many of the free and easy habits that have made the Tahitians famous," and "when men grappled they fought like savage beasts" (79).

In this context, Caudill describes the numerous cruelties of coal mining companies, such as "freeing" the coal camp just as a mine was to be closed. This occurred when the company would announce it was "going out of the real estate business" and sell the company houses to their occupants. Caudill points out that this cynical practice allowed miners to briefly "enjoy the feeling of independence and self-assurance that comes from home ownership" only to find that the house had become worthless and the community economically dead (263). Yet these economic practices are not the end of the story, as we see in Caudill's conclusion. Despite the hyper-exploitation of the area's labor power and natural resources, it was his view that the welfare state had really degraded the people of eastern Kentucky. The mountaineers' "naturally" loose morals were further degraded by the dole's encouragement of "welfare mothers" (287). The welfare system was the final nail in the coffin, destroying "the old fierce pride and . . . spirit of independence." People whose "parents would have starved before they would have asked for charity" were "now in their old age shamelessly [plotting] to 'get by' on public assistance" (350). Despite the heavy toll of "miner's asthma" or black lung, unemployment, and the crumbling infrastructure left by the industry's careless construction, in the end, Caudill concludes, "It is scarcely probable, however, that the monstrous industrial exactions have greatly exceeded those of the pistol, knife and whiskey" (351).

In 2006, Caudill's book was cited in *Coal Hollow*, a book of photojournalism and oral histories and a traveling photography exhibit.[10] *Coal Hollow* is a cautionary tale of the long-term effects of a "take-no-prisoners brand of free enterprise" that has drawn attention to coalfield conditions that are usually invisible in American culture (Light and Light 2006, 2; J. Hill 2007). In the introduction, Melanie Light lists other well-known interventions into coalfield injustice, such as Caudill's, and notes that these have had "no lasting or substantial effects" in improving conditions in the coalfields (Light and Light, 2). What might explain this lack of lasting impact is that many of these interventions portray the region in ways that reaffirm it as a national sacrifice zone. For instance, *Coal Hollow* repeats many of the same tropes that appeared in Harry Caudill's account of fifty years ago. The introduction to the oral history section begins like this:

> It took quite a while before I really let this story into my heart and mind because it seemed very, very far from my life. . . . The people in these photographs looked so poor that if I squinted my eyes, they might have been in Africa or Peru. It seemed to me that life

had always been rough in Appalachia, but this degree of poverty was the ugliest, toughest, poverty I had ever seen. It was too hard. Then Ken brought home some archival panoramic photographs of miners and church groups taken by Rufus Ribble in the 1940s and 1950s. . . . I was shocked to see the difference between the old pictures and the new ones. Those earlier miners looked like rough-and-tumble laborers, but they also looked more alive, as if they had pride in their ability to bring home a paycheck. The families seemed to have some energy and hope in the future. The kids wore clean clothes that matched and fit, and they looked alert and happy. How could conditions have so deteriorated in forty years? Suddenly I started to pay more attention. (109)

One thing to note about this passage is that the two types of photography are very different; the photos in *Coal Hollow* are personal, most frequently with one person or a few close family members in a private setting, at home or in their yard. The old photographs referenced in the text are large wide-angle group shots, taken in public, at church, or at work. By eliding these differences, the passage imputes regional homogeneity and suggests that each set of photographs is representative of coalfield conditions. Interestingly, this narrative identifies the deterioration of conditions as having transpired in the forty years *after* Caudill's account.

The book jacket copy, which was presented as text at the photo exhibit, affirms this dichotomous image of the population (as ideal white citizen-workers or degraded hillbillies): "In Ken Light's poignant images and in their own distinctive voices the residents of Coal Hollow—a fictional composite of the communities the Lights surveyed—reveal *how the intersection of mountain culture with the greed of the coal companies* produced the most powerful economy in the world yet brought crushing poverty to a region of *once-proud people*" (Light and Light 2006, inside front jacket, my emphasis).

These passages bring together several common tropes. The authors reiterate the exceptionalism of white Appalachian poverty by noting that it occurs in a region of once proud people, by implication, unlike the Third World poverty it resembles. They also follow Caudill in suggesting that "the greed of the coal companies" is not enough to produce poverty; greed, in itself, is not dysfunctional; it must have intersected with "mountain culture." But what is this mountain culture? In a classic rendition of the culture of poverty thesis, Jack Weller summarizes thirty-four differences between

Middle-class American	Southern Appalachian
Thoughts of change and progress; expectations of change, usually for the better	Attitudes strongly traditionalistic
Use of ideas, ideals, and abstractions	Use of anecdotes
Strong emphasis on saving and budgeting	No saving or budgeting
Desire and ability to plan ahead carefully	No interest in long-range planning
Recognition of expert opinion	Expert opinion not recognized
Responsibility for family decisions shared by husband and wife	Male-dominated family
Striving for excellence	Leveling tendency in society
Cooperation with doctors, hospitals, and "outsiders"	Fear of doctors, hospitals, those in authority, and the well educated
Use of government and law to achieve goals	Antagonism toward government and law

Selected differences between middle-class American and Southern Appalachian culture (Weller 1965).

the "Middle Class American" and the "Southern Appalachian" in the appendix of his book *Yesterday's People* (Weller 1965). The comparisons are as interesting in terms of their normative expectations for middle-class Americans as they are in their construction of Appalachian difference.

Weller's point was that these mountain characteristics were holding back Appalachians from taking their rightful place in American modernity. For instance, describing company town days, he expresses an unusual opinion about the institution of scrip, which was a form of wage in early coal-mining days that could only be spent at the company store, where prices were higher than elsewhere. The system often led to workers owing their employer money because of the low wages for mining and easy credit available from the company store (Corbin 1981; Eller 1982). Weller clarifies:

> The store was a company store, and bound the miners to it through a system of scrip, which was really an advance on the next days'

money. (This much maligned system began as a service to families and ended by entoiling the unwary, because it made credit so easy to get). (1965, 93)

Of course, the miners were always in need of an advance because their pay was so low. And "entoiling" seems to be a euphemism for debt peonage. According to Weller, because of their isolation and ignorance, unwary coal miners found themselves working to pay off debts. But companies even charged miners for their tools, ensuring a certain amount of debt built in to the job (Corbin 1981, 10). In this functionalist paradigm, miners' low wages, coal companies' exploitative labor practices, and monopoly in retail and housing are not enough to explain the problems of coal camp life.[11] The perpetual indebtedness of miners and their families is not related to the structure of coal mining; they just don't value savings. Similarly, for Caudill, the coal industry's exploitative practices and destruction of the landscape were not enough to explain the economic and psychological depression in eastern Kentucky.

By some kind of environmental and cultural alchemy, Appalachia maintains its unified regional identity in the national imagination, with the entity known as "mountain culture" applied to European immigrants, black migrants, and "native" white hillbillies alike. As in the opening sequence of "Appalachian Emergency Room," the mountains themselves tell you what to expect. The inside cover of the dust jacket of *Coal Hollow* reports, "Ken and Melanie Light traveled hundreds of miles through *rugged, isolated terrain* recording the stories of people whose lives were shaped by coal. . . ." (my emphasis). Semple would agree, as she remarked in 1901, "A glance at a topographic map shows the region to be devoted by nature to isolation and poverty" (1901, 147).

The Role of Disgust in Regional Marginalization

Mountains are uncivilized, dangerous places. The normal standards of American culture seem not to apply. Life in the mountains takes its toll on people, as Semple reported:

The mountain woman . . . at twenty-five looks forty, and at forty, looks twenty years older than her husband. But none of the race are stalwart and healthy. . . . Men, women, and children looked underfed, ill nourished. This is due in part to their scanty, unvaried diet, but more perhaps to the vile cooking. The bread is either half-baked soda biscuits eaten hot, or corn-pone with lumps of saleratus [baking soda] through it. The meat is always swimming in grease, and the eggs are always fried. The effect of this shows, in the adults, in their sallow complexions and spare forms; in the children, in pimples, boils, and sores on their hands and feet. (1901, 152)

There is a common subtext in many representations of Appalachia: disgust. The quote above is quite likely to evoke a gag reflex in the reader, as one imagines the "vile" offerings that, according to that famed mountain hospitality, a visitor would be compelled to share. The disgust is amplified by the author's juxtaposition of the description of the food with the itemization of the people's skin problems. Recurrent invocations of filthy bodies and dwelling places generate an affect of disgust that reinforces a disidentification with the landscape and communities of the mountains. These precognitive invocations of embodied abjection rapidly telegraph difference and separation, influencing how even critical images are read.

Dirt

The introduction to *Coal Hollow* emphasizes the "cultural paradox" of Appalachia: "On the one hand, [Appalachians] have been mythologized ... and acknowledged as a source of much American folklore and culture. On the other, they are seen as shameful 'trailer trash,' ignorant hillbillies" (Light and Light 2006, 1). Similarly, there are "two parallel universes" in West Virginia, a middle-class one with paved driveways and brick houses and another, more ominous one:

> The other West Virginia could be mistaken for a slum in some part of the Third World. Coal camps still line the creeks like peas in the folds of an apron, but they are shrunken and dried out. Dilapidated houses and trailers litter the hollows like piles of waste mixed up with denuded forest, jagged, abandoned swaths of strip mines, and toxic slurry ponds. Raw sewage flows down the creeks of the most beautiful mountains in our country. Clumps of toilet paper still cling to tree roots, left from the last floods. Big cities like Welch and Mullins that once teemed with a hundred thousand people or

more are now cavernous disintegrating masses. Aging and disabled miners, their widows, and a lost generation of people who have never lived in a viable economy are hanging on, passing time in front of the TV or "settin" on the porch. Along with mineral debris, the coal companies left behind human slag. The broken earth and the broken people await reclamation. (4)

This passage vividly describes the deplorable state of the coalfields but also indexes powerful images in the American cultural lexicon: "a slum in some part of the Third World," trailers littering the hollows, toilet paper and sewage in the streams, and finally, the evocative human slag, "settin" on the porch.

A network of associations around Appalachia enhances these phrases with connotations that escape the author's control. These images are saturated with cultural understandings of Appalachian poverty, implying commonsense understandings of what this poverty means, who is at fault, and why. However well meaning, this effort to publicize the tragedy of coal extraction indexes the cultural complex of Appalachian marginalization, only to leave the reader with a familiar sense of disgust.[12]

Like the dusty cowboy of the movie Western, the dirty-faced hillbilly is a common image in depictions of Appalachia. For instance, "hillbilly horror" films specialize in creating a sense of place through filth. This film genre is instrumental in the cultural marginalization of rural spaces through the repetition of certain tropes that center a city subject (Bell 1997). In addition, these images of dangerous "white trash" reaffirm the cultural logic of differential worthiness as they express the tensions of American racial and class politics (Hartigan 2005; Newitz and Wray 1997). Usually the story consists of middle-class urban or suburban people getting lost in a remote rural space where they are attacked by crazed locals; some examples of the genre are *Deliverance, I Spit on Your Grave, 1000 Maniacs, Jeepers Creepers, The Texas Chainsaw Massacre,* and *Wrong Turn.* One of the themes that unite these films is a progressive sense of danger as one leaves civilization behind, which is sometimes symbolized by a loss of cell phone reception, as in *Wrong Turn.* Additionally, the crazed locals are often sexually deviant, as in *Deliverance,* and irrationally violent, sometimes with a psychotic hatred of outsiders.

Wrong Turn generates a thick affect of disgust regarding the mountains and the people who live there.[13] This disgust reinforces the social distance of unmarked viewers from the people and places thus imagined. As with the intro to *SNL*'s "Appalachian Emergency Room," the landscape is not Appalachian; in fact, the film was made in British Columbia. The camera pans over thick woods and broken trees, which indicate a sense of something amiss. The movie is not subtle; we are introduced to "Greenbrier Backcountry, West Virginia" by a couple of rock climbers who are brutally killed by some whooping menace that travels invisibly through the woods like an animal. Then the introductory credits are displayed over a montage of realistic-looking newspaper clippings, phrases, and pictures that tell a story of "West Virginia," a "'Mountain Men' legend," "deformity caused by inbreeding" (next to a picture of a white boy with a severely cleft palate), a cabin, "resistance to pain," "strength," "psychosis," "missing person," "violent outburst," and "mountains" (next to a picture of a campground map). We see cells dividing. We see bloody knives and paths through the woods. Clearly, the only suspense in this movie is when people will die. This extraordinary expository opening montage sets the scene in a way that seems to detract from the movie's thrill factor. We know exactly what will happen and why. However, it does create a strong, seemingly realistic sense of place in which the familiarity of the scene contributes to the pleasures of the film. When I showed the introductory sequence to a sociology class in California, some students remarked that they thought the newspaper clippings were real. One asked me, "Weren't there really some murders on the Appalachian Trail?"[14]

Next we see an attractive young white doctor on his way to a job interview in North Carolina who finds himself caught in a traffic jam. We are further introduced to the place by the raving voice of a preacher on the car radio, denouncing incest. A news bulletin informs us that two college students have been missing since going rock climbing last Monday. Our doctor rashly ignores these warnings. Determined to get to his interview, he braves the backwoods in an increasingly menacing succession of encounters with the rural Other. First, a surly white truck driver with a southern accent insults his hair (too well groomed). The doctor takes off, looking for another route, and finds he has no cell phone reception. Looking for a phone, he stumbles upon a crumbling gas station, manned by a semicomatose old hillbilly in overalls, covered in filth, with a few nasty teeth sticking out of his mouth. Our polite doctor asks, "Excuse me, sir, but do you have a pay phone?" The hillbilly seems to resent the doctor and doesn't offer any help. The only phone is broken and filthy. A dirty old road map hanging on the wall suggests another route, "Bear Mountain Road," and the doctor is on his way, telling the old man, "Take care." The complete indifference, even complicity, of the hillbilly is indicated in his sotto voce response, "You're the one gonna need to take care." Clearly, the local knows what is out there and doesn't care, perhaps even enjoys the knowledge.

After a collision with the SUV of another group of misplaced white city folk, the doctor and these new friends take off on foot to look for help. The movie becomes increasingly (self) referential to the hillbilly horror genre, maximizing both the humor and the familiarity of the story. First they find a road that is not on the map ("That's because you don't have the redneck world atlas"). Then they find a creepy little dark cabin with a tin roof and a low front porch. The cabin is surrounded by a dirt yard and ancient, abandoned cars; the characters express misgivings. When one woman suggests they keep walking, one of the men comments, "What, the next house is going to have a white picket fence?" As they approach the door, the same man says, "I was just thinking, you know, West Virginia, trespassing, not a good combination. . . . I need to remind you of a little movie called *Deliverance*." Nevertheless, they knock on the door, and when no one answers (the inhabitants of the cabin are busy killing the couple who stayed back with the cars), they just walk on in. The cabin is grotesque. Flies are buzzing; something putrid is simmering on the stove. The nauseating remains of a meal are left moldering on the table. The cabin is primitive—there's a wood (or coal) stove, an old Victrola-style record player, and a generator to supply electrical power to the refrigerator and the lightbulbs hanging on wires from the ceiling. As they look around the kitchen, one of the women asks, "OK, who lives here?" The other woman answers, "I don't know, but could you help me find the bathroom?" Her boyfriend replies, "Baby, I think this is the bathroom."

Gradually the mere grossness of the filth is supplanted by more menacing discoveries: jars of organs in the refrigerator, barbed wire in the drawer (used to trip people and flatten tires), and piles of out-of-place sporting goods—things that "obviously" don't belong there, sunglasses, climbing gear, paddles, and toiletry packs. One of the men suspects a cult: "Economically depressed places are breeding grounds for apocalyptic cults." Astonishingly, one of the women uses the bathroom, where the toilet is filthy and there are jars full of slimy dentures on the counter, and they are on their way out when the inhabitants of the house arrive with the body of one of their friends. The city folks hide. The mountain men are bestial;

they communicate by grunts, birdcalls, and crazy laughter. They are not just inbred but apparently cannibalistic as well. After a scene of butchery in the cabin, they fall asleep and the city folks escape, only to find a virtual parking lot full of SUVs behind the cabin. One of the group wonders, "How can they get away with this?" but we already know the answer to this question, having seen what the "innocent" locals are like back at the gas station. The rest of the movie consists of the city folks trying to get away and mostly not succeeding.

Cabins

Movies like *Wrong Turn* can shape experiences of the landscape and its people. These are powerful images, repeated so often that they become naturalized in gut feelings and paranoia. A form of embodied knowledge, this social distance feels real. For instance, after driving up a long, winding mountain road to meet an activist, I passed a group of people sitting on the front porch of a small cabin. I drove nervously on, thinking that can't be the place! My image of environmental activism did not include, at that time, people sitting on the front porch of a mountain cabin—I was actually afraid to talk to them. But they called out to me, and asked where I was headed, and of course, the activist was there. At this point, my vision of the group changed; suddenly, I saw a group of individuals. Before I realized who they were, I was in some sense unable to see them properly. I could only see the ghosts of all the scary hillbillies I'd ever seen on TV.

The activist and his family invited me and my daughter to eat with them. The food was delicious: homemade applesauce, wonderful meatloaf, crispy fried potatoes, and cornbread. The food had been cooked on a coal stove; the cabin had no electricity. I thought—people live like this? In the course of our conversation, it became clear that this was a weekend getaway. In the same way I didn't recognize the activist, the place didn't look like a retreat to me at first.

Along with the sound of banjos playing, the cabin—and other forms of nonstandard housing, like the trailer—are powerful signifiers of place. The cabin conjures an entire mythology in the national imagination. Paul Webb's cartoons that appeared in *Esquire* in the 1940s generated a genre of cartoon humor about Appalachia, with characters like Snuffy Smith and Li'l Abner, which features a distinctive graphic style focusing on floppy hats, overalls, and indolent postures. In Paul Webb's *Esquire* cartoons, the humor frequently focused on male hillbillies and dirt: mules in the kitchen, flies, outhouses on fire. The action is usually located in front of a cabin or out-house, or inside the cabin (Batteau 1990, 129–131). Along with the long beards, bare feet, and scruffy clothes on the hillbillies, the cabin serves to place the cartoons.

The cabin in *Wrong Turn* is a terrifying place, a dark and nasty place. The same kind of menacing housing helps place *Deliverance*.[15] *Deliverance* is the story of four men (Lewis, Ed, Bobby, and Drew) from Atlanta who decide to take a last-chance canoe trip down a mountain river that is soon to be flooded by a hydroelectric dam. Most of the people from the area have already been evacuated as the men approach the head of the river. As in *Wrong Turn,* an increasing sense of danger is created by a series of encounters with the rural Other. In *Deliverance,* the confluence of place and social class is highlighted; the men are middle-class urban southerners confident in their superiority over the poor mountain people. Led by the excessively masculine adventurer Lewis, the men's voyage to the mountains indexes the white American cultural narrative of the frontier as a place to prove oneself and improve oneself (Braun 2003). Similarly, the endangered mountain communities are on the verge of "melting away like so many Indians" (Frost 1899, 115).

The famous "Dueling Banjos" scene takes place when the men stop at a cabin/gas station to hire someone to drive their cars to the end of the river. They think the place might be already deserted, but an old man walks out of the woods and asks if the tourists are from the "power company." This notation of the destructive nature of energy production, and Appalachia as a sacrifice zone, provides a frame for the film. Bobby, later to be raped by a hillbilly, makes fun of the old man and comments to his friend: "Genetic inbreeding. How pathetic." This elitist disgust for the uncivilized mountain people, embodied by the least physically fit member of the party, also reaffirms the film's gendered conception of the mountains as a frontier—these men are too soft to handle the wilderness they will encounter. While they gas up, Drew plays a banjo duet with a strange-looking (inbred) boy sitting on the porch of the cabin, and the old man enthusiastically and unself-consciously dances to the music.

The men then visit a cabin where they see disturbing glimpses of an unimaginable life through a shadowy window—a disabled girl and a diseased old woman sit in blank-faced inactivity. In a trope directly borrowed in *Wrong Turn,* the life lived inside these dark interior cabin spaces is completely Other, and it foreshadows the trouble the men are going to find in the woods. Inside a dark shed, they find the man of the house, with whom Lewis haggles over the payment for driving their cars down the river. The mentally deficient people the men meet are totally identified with nature; the mountain people, like the mountain river, are dangerously wild. Drew's attempt to reach out to the mountain people through music seems to have been in vain; as their canoes pass under a swinging bridge, the boy from the gas station stares at them without recognition. The human community and the river valley it inhabits is soon to be destroyed in the name of progress; given the energy demands of modernity, and the degenerate backwardness of the mountain people, this may be the optimal use of the land. The film suggests that the only way modern society can relate to nature is through total domination; trying to meet nature on its own terms, as the men do, with their canoes and hunting bows, ends by reducing the men to their elemental nature—they must become the same as the savage mountain people or be destroyed.

Along the way downstream, Bobby and Ed are set upon by a pair of hillbillies wearing Civil War–era clothes and carrying ancient muskets. The hillbillies explain to the tourists that they've made a "wrong turn." One of them ties Ed up while the other rapes Bobby. They are about to begin on Ed when Lewis shoots the rapist in the back with an arrow and the second hillbilly flees. The men face a decision on what course of action to take. Drew urges them to turn the matter over to the police. Drew, who wears glasses and believes in communication, represents an overly civilized and emasculated intellect. His sympathy for the mountain people seems to indicate a naive idealism (Armour 1973, 284). Following the advice of Lewis instead, the men bury the rapist in a shallow grave; the hydroelectric dam will soon cover the crime.[16]

They proceed downstream until they run into raging whitewater, where Drew suddenly falls into the water. He appears to have been shot—or maybe he had a stroke—the river was too loud to be sure, and in the ensuing panic the canoes overturn. Lewis comes out of the rapids with a compound fracture. The now disabled Lewis and his remaining two companions, Ed and Bobby, find themselves trapped in a narrow gorge between two sets of rapids. Drew is dead; he emerges from the rapids badly mangled by the river. They decide to act preemptively, and Ed scales the cliff and shoots (with an arrow) a man he presumes is the other would-be rapist. When he brings the body down to Bobby, however, it is clear that he's killed an innocent man. They sink the man's body and the body of their friend. The river has uncivilized the men—they've become like the hillbillies who attacked them, committing acts of irrational violence. Significantly, the most "civilized" of the men is killed. Finally, the men escape back to civilization, where the police investigate their story halfheartedly; soon enough the place and the community will be gone. The titular deliverance of the story in the end seems to be the hydroelectric dam itself, which, the film suggests, is after all the best use of the land.

Bodies

In the opening montage of *Wrong Turn,* the murderous mountain men are explained as genetic mutations caused by inbreeding. Incest is one of the most powerful stories told about Appalachia, as it reaffirms meritocracy and identifies poverty with embodied difference. In his discussion of the alleged culture of poverty in the mountains, Weller opined that some of the conditions there might well be genetic: "One often finds only two or three family names in a valley, and double first cousins are not uncommon. It frequently happens that a girl marries and does not even have to change her last name. One wonders how much this close intermarriage has affected the basic stock of the people of southern Appalachia" (1965, 13).

The popular clothing retailer Abercrombie and Fitch introduced a T-shirt on this theme in 2003: "It's all relative in West Virginia." West Virginia governor Bob Wise asked the company to take the shirts off the shelves; they refused (Dao 2004). Instead, they introduced more T-shirts in the same vein, including "West Virginia: No lifeguard on the gene pool" and another that, ironically, made fun of Kentucky for not having electricity (Associated Press 2004).[17]

Wrong Turn is an extreme example, but this epistemology of disgust for Appalachian bodies and places pervades even more mundane depictions. Anglin discusses a 1994 memoir, written by a doctor, Abraham Verghese, about his time as a "foreign" physician in the Tennessee mountains. She writes:

> There were, according to Verghese, two kinds of people in East Tennessee: "rednecks" and "good ole boys." The former term applied to people whom he characterized as living in "little hollows," in trailers "with no underpinning and dogs all around . . . and little

children playing under[neath]. . . . *That* world was food stamps
and ignorance and rotted teeth and rheumatic fever and a suspicion
of all strangers. (1999, 268, emphasis in original)

"Good ole boys," on the other hand, had "evolved from fighting with the
Indians and feuding with each other to become folk who, as they told you
themselves, would give you the shirt off their back—if you were their friend"
(Verghese in Anglin, 268). The "good ole boys" were different from the
"rednecks" not only because they had "evolved" (from fighting Indians)
but also in their daily habits of consumption: they shop, they have access
to high-tech medical resources, they wear proper clothes. Once again, we
see the "real Americans" and backwards hillbillies side by side. Verghese's
good ole boys and rednecks also recall *Coal Hollow*'s two West Virginias,
the working one, with paved driveways, and the other, the human slag.

Because it seeks to draw attention to an ultramarginalized, usually invisible group of people, *Coal Hollow* does not explore the "functioning universe of employed West Virginians" (Light and Light 2006, 4). However,
this focus entangles the story in a tradition of representing Appalachian
difference that helps construct the region as a sacrifice zone. The photographs document poverty and social exclusion. Many are of people inside
or in front of their homes. These may be decrepit trailers, cabins with plywood walls, or old coal camp houses with makeshift electrical systems.
Many of the titles include the subject's age, as in the photo of Lorraine, on
page 1. This photo was part of the exhibit at Track 16 Gallery in Santa
Monica, California. Other photos in that exhibit included "Bryan at the
Swimming Hole," showing a heavyset boy in white underwear in front of
a mountain creek (10); "Roadkill" (8), a picture of a dead fox in the middle
of a road; and "Ricky, Forty-seven Years Old. Fireco, West Virginia, 2001"
(9). Lorraine is possibly blind in one eye. In another photo (14), she is
shown smiling, toothless, in front of a burning multiflora rosebush.[18] These
images are saturated with layers of cultural connotations that inform how
they are interpreted and what affective responses they generate (Hall 2000;
Coombes 2003). For instance, echoing Semple's commentary of a century ago, Lorraine's age is noted in both titles; at the exhibition, one well-
dressed young woman who had dropped by the gallery early that Saturday
evening remarked, as if on cue, "Fifty-two years old? She looks eighty!"

Ricky's first picture (9) is a close-up of his face. He is shirtless and mostly
toothless; he smiles for the camera. In his second photo in the collection

"Lorraine, fifty-two years old, on the porch." Tommy Creek Hollow, West Virginia, *2000. Reproduced by permission of Ken Light. Photograph and caption copyright Ken Light/*Coal Hollow.

(69), the suggestion of "an empty gaze" in his smile is affirmed (Light and Light 2006, 3). In this picture, Ricky smiles for the camera, with his arm around Eva, who is staring, apparently vacantly, into space. They sit hunched over on a buckled, bare mattress. The room is lit with a bare bulb hanging from the ceiling; the walls are bare slats.

The accompanying text does not specify what images such as the dead fox pictured in "Roadkill" and "Ricky and Eva" teach us about "a government that yields too easily to pressure from industry" (Light and Light, xi).

"Ricky, forty-seven years old, and Eva, forty, husband and wife, in their bedroom." Fireco, West Virginia, 2002. Reproduced by permission of Ken Light. Photograph and caption copyright Ken Light/ Coal Hollow.

There may be a lesson there, but it needs more exposition, especially because the picture facing Ricky and Eva's, titled "Freddy, Fifty-six Years Old, from a Family of Fourteen Children. Tommy Creek Hollow, West Virginia, 2001" (70), communicates a very familiar story, particularly when placed next to the "husband and wife, in their bedroom." Freddy is smiling widely and toothlessly and is apparently blind. His sightless gaze, placed next to the dismal prospect of Ricky and Eva's marriage on the facing page, creates an impenetrable wall of alterity, to the point of the monstrous. This pair of

photographs wordlessly indexes countless images of inbreeding, degradation, and miserable isolation that overwhelm and dilute the book's critique
of the coal industry. Because it summons these affective responses, so ingrained in readings of the region and its problems, the structure of the book
leads away from the political critique signaled by the foreword written by
former secretary of labor Robert Reich, toward the "juxtapolitical," a sentimental story of hardship and suffering that naturalizes the place of Appalachia (Berlant 2008, 10).

Several of the images in *Coal Hollow* only make sense from within a ritualized repetition of signifiers, designating conformity to a genre (Hollywood
2002). At the exhibit at Track 16 Gallery, for instance, some of the more
conventional photographs on display included images of worshipers extend-
ing their arms and praising God (Light and Light, 23), of hooded white
supremacists (26),[19] and of a photograph of a bearded man from the nineteenth century preserved behind glass on a gravestone in "Hatfield Grave.
Hatfield and McCoy Trail, McDowell County, West Virginia, 2002" (34).
Along with the sightless, toothless smiles of Lorraine and Freddy, these pictures effectively place Coal Hollow. Their role as place markers seems to be
the logic for their inclusion, as they don't specifically address the economic
problems that are the book's focus. These images reconfirm the context
by repeating what we already "knew." This categorization implies that
there are important and fundamental differences between these people and
us, the readers, reassuring us that we are safe from what has happened there.

Despite the very real poverty of the coalfields that the photographs
depict, the book itself is a culturally embedded construction. The act of
taking and selecting the photographs is a process of representation that
is overdetermined from the outset. For instance, black people are overrepresented among the poor in several of the counties represented in *Coal
Hollow,* a social fact related to interactions of labor history, geography, and
continuing patterns of social exclusion, among other things. Yet all the photographs in the book are of white people, indicating the power of genre to
minimize complexity (Braun 2003). In the film that was shown at the
exhibit, the narrator remarks that West Virginia is a case study of what happens when corporations have too much power, of capitalism unchecked,
but the "placement" of the subject undermines this goal. The culturally
entrenched narrative of Appalachian difference wordlessly reassures the
reader that it is only the less worthy or the less able who suffer from "capitalism unchecked."

Appalachian Values: When They Are Good They Are Very, Very Good; When They Are Bad They Are Horrid

Because *Coal Hollow* focuses on the poorest coalfield residents, it suggests that West Virginia's parallel universe where people have "cozy brick homes [that] sit on well groomed lawns, with a late model car in a driveway of newly poured asphalt" (Light and Light 2006, 4) is not part of the story of coal. Those West Virginians who live more or less as "normal" Americans cannot be located within this narrative and therefore neither can "we." The book's attention to extreme cases ironically interpellates its unmarked readers—those who do not feel implicated by these images—as *safe* from the kind of economic and environmental disaster it depicts. In this depiction, coal mining looks like someone else's problem; readers are called on to pity the helpless, not to save themselves from the environmental and economic costs of coal.

This is an instance of what Plumwood calls the logic of colonization; in order for Appalachians to be culturally legible *as such,* they must be radically excluded from the everyday world of American life (1993). Consider the role of geographical location in one famous, iconic depiction of the hillbilly, *The Beverly Hillbillies,* a TV show that ran for almost a decade, from 1962 to 1971, and has lived on in syndication ever since. The show was about the Clampetts, a family of hillbillies led by a patriarch, Jed—per the show's theme song, a "poor mountaineer, barely kept his family fed . . ." who accidentally strikes oil and becomes wealthy. Immediately, the Clampetts are out of place in the mountains. As the theme song goes, "Kinfolk said, 'Jed, move away from there'; / Said, 'California is the place you ought to be,' / So they loaded up the truck and moved to Beverly . . . Hills, that is."

Once relocated to California, the Clampetts are comically out of place. Ignorant of modernity, they call the swimming pool the "cement pond," and Granny persists in cooking "victuals" like possum and greens. Along with the comedy of their cultural backwardness, however, the show also regularly used the Clampetts to represent ideal American values, especially in contrast with their greedy and manipulative banker/keeper, Mr. Drysdale. The Clampetts repeatedly stymie thieves and con artists, not through cleverness but through their unimpeachable character. This strength of character is directly descended from frontier battles for land and for property rights, as two episodes from 1967 indicate. In one, Granny learns of a property dispute with a reservation neighboring their oil fields back home and gets ready to defend the mansion against these "Indians" who she imagines are coming to scalp them all.[20] In another, she witnesses a Civil War reenactment and fearlessly prepares to attack General Grant.[21]

The Clampetts' backwardness was indicated on the show by their idiosyncratic speech, clothing, and vehicle of choice, none of which they changed after becoming wealthy. Jed, the patriarch, wore a rope as a belt. Like the rapist hillbillies in *Deliverance*, Granny dressed in nineteenth-century style. The headlights on their 1920s-era pickup appeared to be tied on with rope in publicity photos, and as an online commentator put it, their devotion to this run-down automobile was a testament to their character: "The Clampetts continued to drive [the truck] around town, this unapologetic symbol of their strong-willed independence and immunity to social pressures or the scorn of snobbish neighbors" (Koma 2006). Besides the family's gentle naïveté and the boundless, good-hearted stupidity of Jed's son, Jethro, very few negative stereotypes are on display in *The Beverly Hillbillies*. As the honest backwoods foils to superficial modern society, the Clampetts are clearly good ole boys, not rednecks. In each episode, they bring a touch of "native dignity" to Beverly Hills life (Koma 2006). Kindhearted, generous, and good-natured (as long as one is not an Indian or, presumably, African American), they stand for what is best about rural (white) America. However, the show indicates that this character has to be taken out of its marginalized geographical and economic context to be interesting; the Clampetts have no story until they become rich, and once rich they must leave their extended family and native landscape. This displacement is key. The inspiration for *The Beverly Hillbillies* came when its creator, Paul Henning, was touring Civil War sites in the South in 1959. Reportedly, "He wondered what it would be like to take someone from the rural South in the Civil War era and put them down in the middle of a modern, sophisticated community" (Internet Movie Database 2006). In the end, an imagined rural mountain isolation stands in for "the Civil War era," but the essential element of the comedy remains—rubes out of place. Without the step of displacement, there is only the pathos of rural hardship.

Sacrifice

On January 2, 2006, thirteen miners were trapped in International Coal Group's (ICG) Sago mine in Tallmansville, West Virginia. An explosion immediately killed the fire boss, the crewmember responsible for safety, and the rest of the crew went deeper into the mine to escape the smoke and carbon monoxide and await rescue. They tried to protect themselves by hanging a curtain, and because at least four of the rescuers (portable oxygen tanks that miners are legally required to carry) were not operational they took turns breathing the oxygen they had (McCloy 2006). The story dominated the national media for several days, having been made even more dramatic by initial reports that twelve of the miners who had been trapped underground for nearly two days had miraculously survived. This incident occurred when the body of Terry Helms, the fire boss, was found and word prematurely spread to family members and the media that only one man had died. The other miners' families and friends celebrated joyfully for nearly three hours in a small church near the mine before official notice was sent that all the men but one, Randal McCloy, had died.[22] Suddenly cable news found coal mining interesting; CNN's Nancy Grace[23] turned her moral indignation toward the coal industry: "Why wasn't this mine unionized?" Her outraged query stands practically alone in the Sago mine disaster coverage. In true cable news fashion, the story of corporate greed and negligence, workers' lack of political representation, and the dismal failure of state and federal regulation was preempted by the dramatic twists of miscommunication, false hopes, and a discussion of the personality and character of the principal players. When Bill O'Reilly asked Geraldo Rivera, "How did this happen? Who's the villain here?" the questions referred to the miscommunication that led to three hours of unwarranted celebration among the families of the trapped miners, not the accident itself nor the delay in rescue efforts that may have cost lives.[24] This makes the accident seem routine and focuses on the human drama of the tragedy instead of the push for profit that determined conditions in the mine. As the story of the "emotional roller coaster"[25] of the miscommunication gained traction, questions that initially seemed obvious, about why standing rescue crews had been "phased out" because they were "used too infrequently to justify the expense" and why miners might be afraid to report safety violations at a nonunion mine[26] took second place to a story of a tight-knit, deeply religious community tortured on national television by the dramatic plot twists. As O'Reilly's questions suggest, the apparent callousness of the mine operators, who, in the confusion, allowed the celebrating to continue for several hours even after they realized all twelve miners had not survived, came to stand in for the actual crime; in the past year, the mine had been cited for over one hundred "significant and substantial" safety violations, indicating a "high degree of negligence" (Roddy 2006; Ward 2006a).[27] The "juxtapolitical" drama of the miscommunication allowed the anguish of mining families to become the focus of the national "intimate public"; everyone could cry along with the bereaved widows and children without ever focusing on the structural causes of the deaths (Berlant 2008). At the same time, the familiar story of "coal in their blood" naturalized mining as something coalfield residents choose to do for personal reasons.

Coverage of the disaster often focused on the religious devotion of the miners, their families, and their communities, evoking the Christian nationalism that became prevalent during the Bush administration, especially after the events of September 11, 2001. Larry King asked survivor Randal McCloy's mother and stepfather if the disaster had led them to question their faith.[28] Nancy Grace played footage of an unidentified neighbor declaring, "I'm not kin to any of those men [in the mine], but they're my brothers. You're my brother, and you're my brother, the way I look at it, because I love Christ."[29] On Fox News the conversation focused on the miraculousness of Randal McCloy's survival, given that the oxygen in the rescuers was meant to last an hour, and he somehow survived for forty-one hours. The journalist asked McCloy's brother-in-law, "Any idea how it [the rescuer] lasted?" He responded, "The Lord."[30] The focus on religious faith, and the wonder of McCloy's survival, not only displaced attention from the question of why the oxygen supply was inadequate for almost any serious mining accident, but the theme of Christian piety also helped establish the all-American character of the workers and their families.

The idealization of the disaster victims and their families by the media reflects the superhuman quality of Appalachian virtue, often identified with the landscape; as one reporter put it, "Like the coal itself, grief and heartbreak run through these hills and so does the hope."[31] CNN's Anderson Cooper commented on the "close-knit community" with a "strong spirit" that is "extraordinary in many ways."[32] Family members reported that one of the miners had been reading Scripture the night before the accident, a passage from Corinthians that said, "Love not the things of this world."[33] In a tragically long-standing coal mining tradition, several of the Sago miners wrote letters to their families when they realized they were likely to die. One of the miners' last words for his family were made public: "Tell all, I see them on the other side . . . it wasn't bad, I just went to sleep . . . I love you."[34] Larry King asked Randal McCloy's family to explain why miners "do what they do," and his stepfather replied, "It pays pretty good, and it's the mentality of the firefighter, the policeman, the soldier. Every day they leave the house, there's that chance, but it's what they do."[35] The miners' bravery was a testimony of their love for their families and children, of their willingness to sacrifice. A friend remarked that miners were men that "would give you the shirt off their back."[36]

The Sago mine tragedy brought to national attention the issue of mining safety and the pattern of increasing unenforcement of regulations and loosening of regulatory oversight of the coal industry that coalfield communities have been dealing with for decades. The coal industry has made it very clear to coalfield residents that it considers fatal accidents in mines, like the many fatal road accidents caused by overweight coal trucks, to be just part of the cost of doing business (Burns 2007). After the Sago disaster and several subsequent deaths at other mines, national and state legislators scurried to enact more stringent protections for workers (Ward 2006c). But cable news coverage of the events chose not to focus on the details of state and federal regulations, their ineffective enforcement through nominal fines, and the lack of proportional response to serious infractions (Ward 2006b). Instead, national coverage of the Sago disaster became a sentimental story of dashed hopes and a miraculous recovery. The conventions of the genre—tight-knit communities, simple piety, and hard-edged suffering—allowed the incident to become an allegory of American national character, an iteration of the story of the ideal white Appalachian. As was Jessica Lynch, the miners were portrayed as always ready to sacrifice themselves to provide for their families and the nation. The miners' families and communities justly joined in the sanctifying of the miners on national television; not only did this allow them to express their grief and their love for the deceased, but this is one of the only representations Appalachians can access to escape their stigmatized identity on a national stage. Occupying the narrative of ideal, sacrificing, Christian citizenship allowed West Virginians to portray their communities and the deceased miners in a positive light on the national screen.

This form of collaboration brings with it a whole host of other associations, however. The repeated references to mysticism and coincidence in stories told about the miners' last days and the miraculous survival of the youngest miner index stereotypes of Appalachian traditionalism and irrationality. News reports subtly underlined this stereotypical distrust for authority in their coverage of the miscommunication that led to unwarranted celebration in the church where family members waited for news.

CNN repeated footage of Anderson Cooper asking a woman, fleeing the scene just after the bad news had just been delivered, about fistfighting in the church.[37] Between reports the network replayed footage of a man speaking vaguely threateningly of finding "the person who done that." Dehistoricized by the fuzzy, eternal quality of these representations, the sense of paranoia about the coal industry that permeates the coalfields recalls stories of Appalachian mysticism, isolation, and suspicion of outsiders instead of a rational distrust of an industry whose indecency they have too often experienced.

Consequently, the coverage of the Sago mine disaster reestablished Appalachia as a place forgotten by time and a place naturally devoted to coal mining. Reporters expressed America's surprise that this kind of disaster could "still" happen and wondered what makes coal miners "do what they do." The coverage told an old story of coal mining that is determined by images of underground coal miners and coal-centered communities. This story mostly ignores the reality of coal mining today and how it is changing. ICG, for example, is a specific kind of coal company; instead of envisioning a long-term relationship with the workforce and community, as the traditional coal companies have done, ICG is a new, flexible coal company, seemingly following Massey Energy Company's decentralized corporate model designed to evade organized labor and regulatory oversight (Brisbin 2002; International Coal Group 2008). But the national coverage of the Sago mine disaster was not focused on the kind of questions that would require specialized knowledge, such as the way that companies are using subcontractors and subsidiaries to avoid paying regulatory fines and hide their accumulating infraction records. In an indication of the superficiality required to represent Appalachia in an easily grasped way, MTR was entirely invisible in the coverage, which focused exclusively on underground mining.

A Place Apart

In reality, Appalachia and the coalfields are as complicated a mess of groups, geographies, and problems, with as blurred boundaries and as many connections to other places and people, as any place at all. Their identity as the culturally and economically marginalized places America knows must be constantly reaffirmed and reiterated (Massey 1993). The problems caused by coal extraction are real and have multiple causes and complications;

they require careful analysis, not mythological treatment. Generic representations of Appalachia—stories of sacrificing citizenship and embodied abjection—repeatedly route discussions of the region away from questions about property and national identity, work, and the human relationship to nature toward conventional narratives in which suffering and poverty are their own explanation. These epistemologies of disgust and social distance help create the conditions of possibility for some of the most dangerous environmental exploitation in the United States and the designation of Appalachia as a sacrifice zone.

The pervasiveness of these representations presents a dilemma for Appalachian activists who must represent their cause in the national media. One activist told me of a presentation he gave to a group of college students. After his talk, some of the students came up to him and told him that West Virginia was just like a Third World country. This comparison angered him for many reasons but primarily because it denies the interconnectedness of Appalachia and the environmental destruction going on there with the everyday life of the United States and makes this destruction easier to ignore. Activists repeatedly have to find ways to make people care about the Appalachian Mountains and the communities who live there. As noted in *Coal Hollow*, despite repeated efforts, it is difficult to focus America's attention on Appalachia (Light and Light 2006, 2). This difficulty indicates the cultural distance between the imaginary place of Appalachia and the American nation and the role of representations in (often unwittingly) reproducing it. The cultural logic of colonization creates Appalachia as a forgettable region; it is backgrounded as a place where nothing happens. "Appalachian culture" is constructed as something radically different from normal American life. It is defined by its difference from an unmarked America. The landscape of the region, along with the labor of its residents, is objectified as the means to America's ends, and its destruction is imagined as not affecting the rest of the world. Together, the landscape and the people are homogenized and bifurcated into either a willing sacrifice or a dangerous wilderness (Plumwood 1993).

In 2004, C-Span aired a debate between Julia Bonds, a leader in the anti-MTR movement in West Virginia, and Bill Raney, the president of the West Virginia Coal Association.[38] She described the effects of MTR on communities: increased flooding, slurry spills and valley fills, while he argued that MTR might actually reduce flooding, because, he said, water is less likely to run off something flat like a table than something steep like a wall.

Reflecting the disidentification of "ordinary" Americans from Appalachian problems, the moderator asked Ms. Bonds why people should care about the destruction of mountain culture. She replied, "Our culture is special. Some people call us hillbillies. That's OK, you can call me a hillbilly. Just don't call me an ignorant hillbilly because I'm not ignorant." In this debate, the activist found herself in the position of having to explain to a national audience why they should care. In order to do so, she relied on the figure of the hillbilly—that image places the problem for the audience; it makes the place legible. But she also faced the negative side of this placement. She is only too aware of how Appalachian marginalization affects the way that both local residents and others perceive the region. In talks about MTR at coalfield schools, she always closes with the line "Never be ashamed of who you are and where you came from."

CHAPTER 2

Men Moving Mountains: Coal Mining Masculinities and Mountaintop Removal

Mining Is Not Naturally/Necessarily Masculine

As the Sago mine disaster coverage in the national media made clear, coal mining is symbolically associated with specific forms of masculinity and community. These associations have multiple and often contradictory valences regionally and nationally. Many of the symbolic forms of masculinity that mining communities take pride in are devalued in the national discourse and associated with backwardness. For instance, coal mining masculinity is articulated with traditional gender relations in the family; a coal miner is a provider, but this form of masculinity is frequently portrayed as threatening to or oppressive of women when it is depicted in the national media. Coal mining also indexes physical strength and hardworking masculine bodies, but these images also invoke symbolic racializing discourses of abject labor and an excessive closeness to nature. Similarly, coal mining is often associated with the past, in its sepia-toned images of company towns, lunch buckets, and canaries, but it is also a crucial site of modernity, signifying technical skill, innovation, and the mastery of nature.

In all these multiple articulations, coal mining is culturally constructed as a hypermasculine industry, despite a long history of women working as coal miners (Beckwith 2001). During the eighteenth and nineteenth centuries, women and children were essential workers in the coal mines of Europe, digging coal, hauling coal cars out of the mine, holding doors open to prevent explosions. The consolidation of the bourgeois ideology of separate spheres led to women's eventual exclusion from European mines (Halsall 1998; Hilden 1993). The masculinity of mining is a historical accomplishment that is continually challenged by the presence of women in and around mining operations (Hinton et al. 2003; M. Moore 1996; Norris and Cyprès 1996; Tallichet et al. 2003). Invented superstitions about women causing bad luck in mining have helped exclude women (M. Moore 1996, xxvi; Hinton et al. 2003).

In the United States today, relatively few women work in the mines. Beginning in the 1970s, affirmative action programs attempted to break the barriers that kept women out of high-wage jobs dominated by men (Norris and Cyprès 1996; M. Moore 1996). As Sandy, a retired white woman miner, told me, at that time the union was interested in getting more women hired. At the mine where she applied for a job, the mine supervisor told her, "If I have to have a woman, it might as well be you." Sandy says that today there are still some women miners out there, but fewer than before. Sandy identified the decline in women miners with the decline in union mines. When I asked her why there were fewer women miners today, she said, "Really and truthfully, I think the union hired a lot more than nonunion places. I don't know why . . . probably it's like equal rights. We got our jobs through equal rights."

Sandy said that although she faced a little resistance at first from the men she worked with, being a woman in a field dominated by men didn't cause her many problems at work. However, she also showed me a picture of the machine she usually operated in her capacity as a rock duster; her coworkers had painted it light pink as a joke. Women coal miners are apparently rare enough for them to recognize and notice each other's existence. Sandy passed another woman miner regularly as she drove home from work and the other woman was on her way in. They were not acquainted but waved to each other in their mutual recognition. The stark visibility of these women highlights the unmarked masculinity of coal mining. Other miners I interviewed recounted stories of women they used to work with, or reports of a woman working in some other mine, or on another crew, that seem to reflect a reassertion of coal mining as a masculine domain.

As a young woman, Sandy made the decision to become a miner in part because her father was a miner. It was a decision she made for economic reasons but in the context of certain gendered expectations. Her experience demonstrates how people's participation in the economy is saturated with cultural associations of gender and sexuality. She told me the advice her father gave her after she took her forty hours' training and had been offered a job: "If you don't think you're ever going to marry . . . it's good money. But the only thing I'm going to tell you is, they give you a job for a purpose. They need you there. They don't hire people just to give them a job. And you're expected to work every day." Sandy added, "And I went to work." This fatherly advice illustrates a few of the ways mining is understood culturally in the coalfields. Economically, it provides a breadwinner wage. As a woman, Sandy could potentially access this wage through het-erosexual marriage or through employment in a "man's" field. Her father's advice indicates he saw these alternatives as mutually exclusive. Addition-ally, he felt it was necessary to warn Sandy of the kind of job she was get-ting into; the mining company expected its workers to *work,* this advice suggests, in a way he thought most women were unaccustomed to. Here, work is identified with mining, making economic activity in the domestic, informal, or service sectors seem merely incidental.

Besides highlighting the everyday assumption that coal mining is a mas-culine job, Sandy's narrative of her decision to become a miner and her father's reaction to this decision suggests how the stories people tell about mining work to embed it in a cultural context. Sandy's story signals two of the most important ways that mining, particularly underground mining, has made cultural sense in the coalfields for most of the twentieth century. Thanks to organized labor, mining has provided a relatively high wage, and a miner is understood locally as a provider, a family man. In other words, one way that mining is understood is relationally—a miner exists in a cer-tain relationship to a family of dependents. But the story also indicates the embodied experience of mining. Miners are not just family men; they are extraordinarily tough guys. Especially in underground mines, local stories suggest, miners daily face physical challenges that would overcome the average man. The work they do is *real work;* they have to work hard to keep up with the demands of the company for production and the demands of the nation for energy. The often hostile reaction of men miners and their communities to women miners indicates the cultural significance of min-ing as well as its economic weight. Women miners don't need men; they are independent and empowered. What's more, in a national cultural set-ting that devalues women and things associated with femininity, if a woman can mine coal, it might no longer seem such an incredible feat; as one woman miner commented, "When women went underground the men's hero identity was kind of destroyed" (M. Moore 1996, 151).

Interestingly, Sandy, who worked underground, did not really say much about the third story that circulates in the coalfields about mining; how-ever, in a sense, it is inherent in her narrative. Coal mining's cultural sig-nificance is not just about men in relationships and men's bodies at work. It is also about technology and progress. This third story, of the modern man, places coal mining in a larger cultural context, as part of the history of making the modern American nation and as a site of the progress narrative of modernity itself. Coal not only fuels economic expansion but market-driven technological innovation also reworks coal mining; increasing levels of production throughout the twentieth century have been driven by developments in mining technology. Ironically, the technological power inherent in MTR threatens the existence of the family man and overshadows the tough guy. Modernization is an ideologically central narrative of American national identity in which civil and economic rights are articulated with technological innovation and free-market competition. This process now seems poised to destroy coalfield communities and the very mountains themselves.

Some of Mining's Local Importance Stems from Its Symbolic Association with White Masculinity

In spite of the existence of women miners, mining and miners are overwhelmingly coded as masculine and white in the culture of southern West Virginia. This is evident in throwaway comments about getting "the boys" out before blasting, or the habit of referring to the employees in a mine as "the men." Additionally, black miners seem to have largely left the mines, where their position was never quite as secure as that of white men (Trotter 1990). As Ben, a white underground miner, put it:

> I'm sure there's some out around, but . . . I don't know where they're at. . . . I did work at the union [office] downtown, and there were several black men working [in the mines]. Good workers. . . . Years ago, I stayed up here [in this hollow] and when they had the mines up here, there were a whole lot of [black miners]. We got laid off and a lot of them retired. I mean, a lot of them just filed for disability or whatever. . . . They ain't working no more.

Ben's assessment of the decline in the number of black miners demonstrates the articulations between mining, working, and white masculinity. Suggesting that black miners might have decided to go on disability rather than work disregards the extent to which layoffs and mine closures narrowed the field of employment in mining.[1] Graham, an African American union organizer and former miner, explained how mining companies discriminated against both African Americans and white women through techniques like placing them all together in mines that were scheduled to be closed, then laying them all off. The enduring masculinity and whiteness of mining are thus the results of the historically gender- and racially segmented labor market in the United States that created a "labor aristocracy" of white men (Mink 1990; Roediger 1991; Valocchi 1994). Additionally, finding a mining job can depend on using social networks, and given the prevalence of gender stereotypes and residential segregation both white women and African Americans are most likely at a disadvantage in such a process.

From the 1950s through the 1970s, the United Mine Workers of America called frequent strikes and hammered out strict labor contracts with the Bituminous Coal Producers Association that were published in annual "National Bituminous Coal Wage Agreements," or "bibles." Since the 1980s, however, antiunion corporations like Massey Energy Company have eroded the power of the union in West Virginia, for instance by not participating in the association's negotiation with the union, by forcing the union to allow coal to be processed in nonunion processing plants, and by forcing the institution of partial strikes (Brisbin 2002). According to some coalfield residents, these steps have broken the union's hold on the coal industry in West Virginia as effectively as right-to-work laws have done in other states.

Graham suggested the union's problem was that it got too comfortable. Miners got used to good health care and good money, stopped going to meetings, and forgot their history. One of the major factors in the union's loss of power might be the simple reduction in the number of workers necessary to mine coal. In 2003, coal production was roughly equal to what it was in the 1950s, but with one-eighth as many workers (U.S. Department of Labor 2005; West Virginia Coal Association 2008). Today, Graham often hears from miners who say "We need a union" but are afraid to give him their names. As he put it, miners will have to build support for the union from within. But many are scared and feel only too easily replaced. Jesse, who works on a Massey mountaintop mine, said that Massey CEO Don Blankenship thinks truck drivers "are a dime a dozen" and that a recent vote for the union at his mine had failed (to Jesse's regret). Only about 40 percent of West Virginian miners belonged to the union in 1997, and the number has only decreased since then (Vollers 1999). This could change if Congress enacts the Employee Free Choice Act that will allow employees to form a union by majority sign-up. Currently, Massey Energy Company reportedly screens employees for union sympathies, and several Massey employees I interviewed voiced the (company) opinion that unions were only useful for the lazy.

Despite reports that companies are desperate for experienced workers, the working conditions that obtain in nonunion mines are in decline. Several of my respondents cited cases of workers being blacklisted and companies hiring from out of state. In any case, the number of workers employed in mining in West Virginia has declined to about a third of the number employed in health care and social assistance (Workforce West Virginia 2005). Nevertheless, so far there is no other type of work poised to replace coal mining as the defining type of economic activity in a region known as "the coalfields." Even though there are fewer and fewer jobs in mining, and these jobs are less substantial than they used to be, there is a dissonance between this local economic reality and the level of social and cultural dominance accorded to the coal industry.[2] The historical dominance of the coal industry in the region is reflected in the popular phrase "King Coal." As its feudal overtones suggest, the phrase reflects a degree of cultural significance that isn't determined by economic rationality. Coal mining is thus a rich site for the consideration of the coproduction of economic and cultural forms.

Three Figures of Mining Masculinity: The Family Man, The Tough Guy, and the Modern Man

Although what coal mining means has varied historically, in the United States today much of mining's significance is heteronormative, reflecting a version of human social life and sexuality based on a dichotomous, hierarchical, and polarized construction of gender. Theorizations of gender accountability, or of gender as a regulatory fiction that reads all performances through the lens of heteronormativity (Butler 1990; Lorber 1993; West and Zimmerman 1987), have been critiqued for their relative emphasis on discipline as compared to transformation. Lucal has suggested thinking in terms of stretching the categories; making the concept "woman" large enough, for instance, to encompass what are now read as masculine performances (1999). Risman has similarly noted that it is unhelpful to "label whatever a group of boys or men do as a kind of masculinity" (2009, 83). Because it is based in a context of binary gender, the "doing gender" model can also foreclose consideration of the potential multiplicity of gender formations (Halberstram 2006). These disruptions and eruptions of multiplicity are always in tension with heteronormativity. However, as a technology of hegemony, binary gender performs not only as a way of disciplining individual bodies but as a traveling discourse that is relative to

context and that works in productive ways with other cultural forms unrelated to biological sex. Performances are not read as masculine or feminine
because a male or female body is (necessarily) enacting them but because
of the articulations between certain gendered positions and attitudes and
the discursive configurations of modern American national culture (Cohn
1993). Masculine hegemony is maintained by valorized forms of masculinity that are involved in the construction of "the real" in powerful sites
across the social world (Rowbotham 1974).

The articulation between mining and masculinity is accomplished in
part through stories that circulate in the coalfields, constructing the local
understanding of coal mining. These semiotic figurations of mining are
"performative images that can be inhabited" (Haraway, in Hartigan 2005,
16). Paying attention to this "figural play" allows us to access the "vast reser-
voirs of signification that animate" cultural forms (Hartigan 2005, 16). The
themes of these stories coalesce around three interconnected figurations
of mining masculinity. By analytically separating the figures, it is possible
to discern transformations and consistencies in the salience and content
of coal mining masculinity. These are hegemony-building stories that help
construct a cultural context in which the coal industry makes sense. Coal
mining is overdetermined, however. These stories do not constitute a closed
cultural system; rather they are porous and contingent; they are part of a
web of signification, fraught with internal contradictions, that requires constant reiteration. Although these mining stories operate as tropes that allow
an individual's story to make sense, history and individual creativity transform these stories as they are retold, often in unpredictable ways. In the
narratives of coal industry supporters, they are frequently told in a way
that helps create a cultural logic for MTR. They are also rearticulated in
interesting ways by anti-MTR activists, who strategically articulate their
positions drawing on the cultural materials available, in what Sandoval has
called the methodology of the oppressed, or what Tsing has called collaboration (Sandoval 2000; Slack 1996; Tsing 2005).

These figurations of mining are intertwined, but each may be more or less
salient in different contexts. The family man is a breadwinner, a provider,
with proper adult moral status. Always already in a heterosexual marriage,
with children, this figure represents both the ideal American working-class
citizen and a conservative, potentially sexist masculinity. He is symbolically
white because the racially segmented labor market has historically reserved
the family wage as a privilege of white men, which has helped ensure their

status as national citizens (Gordon 1994; Roediger 1991; Valocchi 1994). The tough guy reflects the bravery and physical strength required by mining and offers a way for miners and their community to counter the stereotype of the dumb or lazy hillbilly while also signaling the abjection inherent in dangerous, other-directed, physical labor.

Because of the ways that MTR and other mechanized mining techniques have changed the day-to-day reality of coal mining, both in terms of the lived experience and the social importance of the work to families and communities, the family man and tough guy figures are being emptied of their content, which does not necessarily reduce their cultural significance. But as mining technology becomes more spectacular, the third figure, the modern man, is helping to make sense of MTR. The modern man exemplifies both the modern technological domination of nature and the dangerous hubris of this effort. This final figure is becoming more prominent as MTR puts coal's mark on the land in a way that was impossible to imagine in an era of underground mining.

The Family Man

In the southern West Virginia coalfields local common sense is that coal really is king, and as the coal industry goes so goes the rest of the economy. This domination works in more ways than one—not only is the coal industry one of the major political forces in the area, but its history is embodied in the towns and communities and its practices still play a determining role in the local economy. The coal industry and other corporate landowners have in this sense created the place; their market imperatives have added cultural meaning to the natural features of the region, a meaning that has the power to shape the details of residents' lives.

The institution of the family wage, exemplified in jobs like coal mining, has contributed to the reproduction of a sharply imbalanced and segregated labor market for the working class. This imbalance in turn helps reproduce the dependence of the coalfield economy on coal (Barry 2001). Since the onset of industrialization, and through much of the twentieth century, the segregated labor market created the ideal—and often the reality, for many white Americans—of man the provider for a largely dependent wife and children. This is a commonplace not just in West Virginia of course but everywhere in the United States. Employment for white women continues to be constructed as a choice, not necessarily related to their femininity, whereas white masculinity has been intrinsically connected to employment and providership (Collier 1995; Connell 1993; Lupton 2006; Willott and Griffin 1997).

Historically, mining town economies were very homogeneous, with most of the men working either as miners or managers for the company or in some mining support capacity. The company supplied everything: housing, shopping, schools, doctors, and so on. Company houses were reserved for the families of employees. This arrangement complicated the relationship between the worker and the company because not only the worker but his whole family, the whole community, depended on a continued relation to the company (Corbin 1981; Eller 1982). By constructing living quarters for their workers, the coal companies prescribed a particular heterosexual lifestyle for their employees. Thus the companies' control over workers and their families extended from the overtly political—union activism, for instance, could get a family evicted—into the personal. If continued occupancy of the family home depended on the employment of the father/husband, a woman's choices in determining her own life and the lives of her children were also limited. Since women were not usually considered employable as miners, their employment options were limited to those auxiliary services needed in the company towns. This low-wage work and the extensive unofficial economic activity of women are rendered invisible in the discourse of the coal camp and the family wage.

As the industry's need for workers declines, the interplay of the economic and cultural significance of mining creates both an economic vacuum and, for some, a resistance to imagining other ways of organizing the economy. White masculinity is also articulated with mining on a more symbolic level. Among its multiplicity of often contradictory meanings, mining provides a testing ground for young men's morality. In a mining community, some of what working as a miner does for a man—even a boy right out of high school—is to enable him to get married, to start a family, to become socially established, with a house, a truck, an ATV; these economic activities are all saturated with moral significance in a heteronormative culture that values market participation and a standard level of familial consumption (McNeill 2010; Probyn 1998). As Laura, a white educator and the daughter of a coal miner, put it, when you get a job in a mine (or marry a miner), "you've made it." She pointed out that this life choice was a quick, socially sanctioned path to adulthood for high school graduates. And when work is as repetitive and physically challenging as mining is, perhaps some

of the satisfaction of it comes from the lifestyle it has historically allowed.
She told me that her father quit school in the eighth grade, "and yet he was
able to raise two daughters, he's got a home up there, it's paid for, you know,
he's got a truck in the garage."

Paul, a white former coal miner who now owns a successful small business, described to me with pride how he provided for his family through
mining: "Me and my wife got married in '73, and we bought that [house]
in '73 for fifty-four hundred dollars. A four-room coal company house that
was in terrible shape. Four rooms. Somebody put a little old bathroom on
the back, a little old porch. Cockroaches running everywhere. So I started
tearing off, cleaning up one thing and another in the early seventies, and
we built a nice house, and then I built a garage separate from it. . . . I raised
my family there." In our conversation, Paul portrayed himself mainly as a
provider for his family, in ways that demonstrate the moral valences of this
role. As the owner of a small business, he was proud of his ability to provide luxuries for his wife: "She drives around in a new vehicle. I keep her
in a new vehicle, and she never has breakdowns."

Another white retired miner, Sam, told me, "We lived good from a coal
mine." His wife, Marie, also white, echoed this sentiment. She voiced the
common opinion, "Anybody who works for the coal industry is going to
be for it, because it provides a good living." She added, "I've gotten a good
living out of the coal industry." Indicating this, Sam reported that two of
their three children had graduated from college and the other chose not to
go. The twist in this family's story is this: Marie had worked at a government job, long enough to retire with a good pension. Sam retired from the
mines with a pension and insurance, but then his former employer declared
bankruptcy. Massey Energy Company bought a mining complex from the
bankrupt company, and a federal bankruptcy court voided the UMWA contract that had guaranteed those workers lifetime health care. Along with
the company's current employees, most of whom lost their jobs, around
four thousand retirees and their families lost their health insurance. Sam
told me about this and said he was one of the lucky ones because his wife
had a pension with health insurance from the government that would cover
him as well. But both he and his wife described their lifestyle as dependent
on the coal industry. Of course, his wages were most likely higher than
hers, but her work had been invisible in the conversation until the bankruptcy came up.

The habitual focus on mining as the only work that matters in the coal-fields continuously erases the paid and unpaid work that women do to support their families and consequently affirms the community's identification with coal. This largely accords with a dominant national culture that naturalizes the heterosexual provider/caretaker family structure and with the evangelical Christian ideology of what have been called complementary gender roles (A. Smith 2008, 127; McNeill 2010). In a narrative space where coal mining is the defining activity of the region, women's paid work becomes nearly unrepresentable.[3] Additionally, much of the work that women do and have done in the coalfields is in the informal economy, or in the private sphere, so is not counted as economic activity. Nonetheless, historically women were very useful to the coal industry. Their work supplemented miners' wages and cushioned families and communities from the economic downturns of the industry. Women gardened, raised livestock, put up preserves, ran informal businesses, and produced goods like clothing, household accessories, and furnishings (Greene 1990; Pudup 1990).

The provider/caretaker family structure is currently reinforced through the increasingly stringent working conditions in nonunion mines. As such, these conditions run counter to the contemporary ideal of involved fathering (Wall and Arnold 2007). While this ideal has not typically been actualized in equally shared parenting, hegemonic masculinity[4] has shifted toward the expectation of a nurturing father (Craig 2006; Kane 2006). Ed, a white underground miner, commented, "Back in the day, a father's role in the family was pretty much to go to work and come home." But, he added, "I don't think I'm just a 'go to work and come home' dad. I do spend time with the kids." His wife, Leann, affirmed, "I do want him to be involved, and he doesn't go without changing diapers." Similarly, Pete, a white underground miner, said he was lucky to be on the day shift. "I refused to work the evening shift. If I have to go to the evening shift, I will quit and go somewhere else … 'cause I can't miss out. My dad was on the evening shift the whole time I was growing up. I never seen him. We had no relationship." Added Ella, his wife, "Which, I guess most kids are like this, but if he calls and gets a hold of his daddy, he's like, 'Hi, Dad, How are you doing? Let me talk to Mom.'"

This represents a change from Paul and Laura's accounts, quoted above, where raising a family, for a father, meant bringing home a paycheck. But round-the-clock work schedules and seventy-hour workweeks make this new fatherhood ideal difficult to achieve. Several miners' wives said if their husbands couldn't get on the day shift, they preferred the "hoot owl," or midnight, shift, because it allowed fathers to participate in the family's evening life. Rhonda, a white educator married to an underground miner, reported feeling somewhat resentful that her husband's evening shift offered him more "alone" time than she had and left her by herself with the kids every morning and evening. As Jesse, who was working on a Massey MTR mine, told me tearfully during a telephone interview, his two older children knew how to ride bikes, but his youngest didn't; he hadn't been able to teach her because of the hours he was working on the night shift.[5] In order to get on the day shift, he said, "You have to have Massey running through your veins, not blood." Nonetheless, he refused to work overtime, rushing home for an hour's nap before walking his youngest child to school.

Despite this new ideal of involved fatherhood, women (even those who work outside the home) are still held primarily responsible for parenting and domestic labor, and a few miners' wives emphasized that the house and children were their territory. Ed remarked of his wife, "She won't let me cook." Although Leann encouraged Ed to be an active parent, she said, "My job is to be involved in the school, know what's going on, talking to the teacher. That's not his job. It's his job when it gets out of hand for me." Frances, a white woman married to a surface miner, compared miners to doctors: "If you are a good doctor and trying to get a practice, a lot of times, your wife will stay home just to support you. . . . I can't imagine [working outside the home] and trying to focus on that and focus on this."

The Tough Guy

Sam described in detail what it was like to work underground in a twenty-six-inch coal seam. He said he has scars on his back from the bolts holding up the roof, and that he did a lot of duckwalking during his eight-hour shift. He couldn't sit down in the narrow coal seam; he had to lie on his side to eat lunch, unless a rock had fallen off the roof and made a high place for his head. Nowadays, he said, it's easier, because the workers can lie down to drive the machinery. When he came home from his eight-hour shifts in the mine, he said, the only thing he could do was recuperate. This kind of work, with its incredible physical challenges, seemed to earn him the right to rest on the couch while someone else looked after the kids. This was Frances's view of her marriage. She told me, "I try to do everything. . . . I don't mean that in a subservial [sic] wife sort of way. I say that . . . he's making unbelievable money, and when he comes home, I want him to be able to relax. . . . I think it is very hard for a coal miner when their wife works because . . . you know, he is really, really tired. And he'll come in and try to do something crazy a lot of times—mow the grass or something. I try to make sure that it's done, so he can't."

Beyond cementing his place as a provider, this work not only requires the family man to be a tough guy but to take pleasure in the difficulty of the work; local common sense is that you need to be very strong and brave to face, or perhaps enjoy, that daily shift in the mines. Frances commented, "I'm not one of the people that say things like that, but he loves to run coal. . . . He likes high pressure: 'Run coal, you gotta run coal.'" When I asked Tommy, a white underground miner, what he enjoyed about mining, he replied, "I'd almost say the risk. And there is a rush to it, you know?"

Underground coal miners occupy a contradictory place of representing in some degree the abject body of labor—a body bent by working in narrow coal seams, at risk of permanent damage by black lung, cave-ins, and explosions, and historically treated as cheaper than a mule by the coal company while at the same time also representing a pinnacle of masculine strength and bravery. Cohen argues that white masculinity functions to hide the real face of labor: "Labor aristocracies elaborated their own set of racialized distinctions to set themselves apart—as skilled men with regular family wages—from the world of the . . . unskilled and . . . poor," and these distinctions were generated at a time when "working and living conditions were conspicuously dirty, dangerous and degrading" (Cohen 1997, 253). Mining especially fits this description. "From just these ranks has arisen the counterclaim that labor constitutes the backbone of the nation . . ." (248). Working-class men's laboring bodies are disciplined by the idea that their "heavy work proves [their] masculinity" and by wage arrangements that convince them they are a "'high-priced man' . . . capable of heavy masculinized work" (Bahnisch 2000, 63).

Mining is hard work but as Sam told me there are a lot of people who love to work underground. Laura remembered that her father "liked the equipment. He was proud of himself. It proved he was a man." Working underground is understood to be especially challenging, so if someone actually enjoys it that indicates an unusual degree of toughness that coalfield communities take pride in. Several commented, "It takes a special breed to go underground." Accordingly, the miner's laboring body is integral to representations of underground mining. For instance, the cover of Barbara

Freese's 2003 book on coal mining represents the industry with the iconic dirty-faced miner, in this case, a boy (Freese 2003). This image of an underground miner, face covered in coal dust, is the ultimate representation of underground mining; the miner is usually white, grim faced, and exhausted, sometimes markedly young or old, never smiling. Considered in the context of popular images of Appalachian masculinity, which are often monstrous, such as the violent feuding hillbillies of Hatfield and McCoy legend, the inbred mountain men of *Wrong Turn,* or the menacing vacancy of the "human slag" in *Coal Hollow,* it is easy to see how the heroics of mining masculinity can become redemptive (Bell 1997).

One difference between underground and surface mining seems to be the level of personal heroics as opposed to impersonal technical power. As Steve, a white West Virginia Department of Environmental Protection (DEP) officer, said, "It used to be more prevalent than it is now, but you [used to be able to] stop in at GoMart [during a shift change] and hear a conversation between two guys, one got off work, you know, he's black and everything, the other one's going to work, [and] you know, how'd you do? Oh man, we did great, we run X amount of tons."

This story of the heroics of mining is common in the coalfields. For instance, Fred, a black retired underground miner, told me about how he and his friends used to go on vacation to New York City and "brag about the mines." Nathan, a white underground miner, explained the different tasks and equipment used in a mine. The roof bolter's job is to secure the top of the mine section that has just been mined, before the rest of the crew can continue mining. "That is one of the most dangerous jobs in the mine. But these guys that do that, they love their job. The guys that run the roof bolters, they say they love that. . . . They know their job, and they can tell if a mountain is shifting, or they can feel it move, or they see loose rock, or they got that sixth sense when to run, you know?" Randall, a retired white underground miner, remarked, "Underground is really exciting work. . . . You got big equipment. And it's noisy, it's dusty, it's dangerous. But you know, when you take a piece of . . . equipment, a miner, and you turn it on . . . it cuts through coal like a hot knife through butter, you know?"

However, these stories of mining emphasizing the heroism of the job and the excitement of running coal were occasionally contradicted in my interviews. Pete, a white roof bolter, reported that the job was "about the same as anything else. It gets old." In our conversation, Pete's wife, Ella, had just recounted an incident that occurred when Pete was working underground.

He had placed his equipment in his work site. "He told me that nature called. I always thought that they'd come outside when they have to go to the bathroom, but they don't. [*Laughter.*] They make little makeshift things where the fan can take . . . the odor on out, but he said that nature called one morning, and he said it never ever calls that early. So he takes care of business, and he heard this horrific noise. So he gets his stuff back together . . . and runs over and his bolter, his piece of equipment, was covered in rock."[6]

In one sense, then, Pete's subsequent characterization of mining as "about the same as anything else" could be interpreted as courageously downplaying the danger of his job. However, his comment also reflects the fact that mining is repetitive, physically taxing work. As Denise, a white activist and spouse of an underground miner, put it, "Coal mining is a hard job. And it does drag people down. . . . Especially when they are working them fourteen, fifteen-hour shifts. I can go up the road every day . . . seeing people, miners, pulled off the side of the road, asleep in their vehicles. . . . I mean it's working them so much that they don't even have the energy enough to drive home."

What coal mining means is situational and context dependent. The multiple valences of coal mining were apparent in Ed's story of becoming a miner. He was not from a mining family, and when he announced his intention to become a miner, Leann told me, his mother said, "You gotta find something else to do. Here, let me find you another job. . . . Just don't work in the mines." At another point in the conversation, however, Ed reported that the romance of mining is alive and well in the coalfields, where "there's a lot [of people] that pretend they're coal miners." When I asked what he meant by that, he continued. "There's a lot of people that just put the coal mine stickers on their truck. And there's a lot of people working the strip mine who pretend, who will go out on the weekend . . . completely, everything in stripes. The [miner's] pants have the stripes on it, shirts, jacket, stripes on it. . . . A lot of coal miners wear their stripes out in town . . . so people can see they're coal miners."[7]

It is also instructive to consider the differences in Sam's and Sandy's accounts of working underground. Sam's story focused on the awesome physical difficulties he faced as an underground miner, how after each shift he came home to recover from working in a twenty-six-inch coal seam in an era with less mechanization. Sandy, however, minimized the work she did as a miner. "I started out like on general labor, which is making belt moves and shoveling belts and all that, you know, and then in the later years, I rock dusted 90 percent of the time and ran a coal hauler, which [means I] took the coal to the belt line and brought it outside the mine. . . . I mean coal mining is brute labor. [But] if you're on setup crews and things like that, if you're working on a coal crew, and running a buggy or something like that, it's not bad. You're just on your big buggy and go to the miner and get your coal and go around and take it to the feeder and then go back and get another load. . . . You could [do it] too. If you could run a four-wheeler, a go-cart, I mean it's not like that; it's a lot bigger. But you could, yeah, you could. If you can drive a car, you could drive a buggy." Sam commented on women miners this way: "Some can do just about anything. If they can drive a car, they can drive a piece of equipment."

The Modern Man

Today, mountaintop mines are becoming as common as underground mines; about 40 percent of West Virginia coal comes from surface mines (West Virginia Coal Association 2008). Underground mines are highly mechanized as well, using continuous or longwall mining techniques.[8] With the increase in technology, the masculine symbolism of mining has become more impersonal, more technological, and more in tune with a modernizing ideal. For some West Virginians, this modernizing industry seems to offer a chance for the state to escape its dusty coal miner and backward hillbilly image. In all these entanglements, MTR is constructed as a masculine technology of control against irrational and feminine-coded Appalachian nature (Batteau 1990; Garlick 2003; Merchant 1980). By creating flat land, it allows West Virginia to overcome its crooked geography and become more like other parts of the United States. A common remark about reclaimed land is, "You wouldn't know you were in West Virginia!" MTR represents an example of the modern ideal of dominating and improving nature; some people consider the flattops an improvement on the original mountainous topography because they (theoretically) provide flat land for economic development. This is normalized through comparisons to other types of development. As Pete put it, "Strip mines, strip malls." Steve commented, "We are restructuring the landscape, in a variety of ways, as men and women, to suit ourselves and what we want, and mountaintop mining is simply one of those, like roads and malls and housing developments."

The language of coal mining has been marked by a strong identification of the worker with the machinery. In underground mines, both are called "miners." But in MTR, the machinery overtakes the worker (and presents the greatest danger). In some cases, these surface workers are no longer even considered miners; some underground miners dismissively call them "dirt diggers." In MTR, the iconic piece of equipment is the dragline, an enormous earthmover with a shovel as big as a house. Long after the mine was closed, an enormous dragline, known as "Big John," too large to move, loomed over the abandoned MTR site in the town of Blair. But even without the dragline, which is too cumbersome for some of the steep mountainous terrain of West Virginia, moving mountains takes giant equipment.

In her study of French farming, Saugères notes that the tractor, which replaces the farmer's physical effort with technological force, is both a source of power and a symbol of masculine strength. Men control the tractor's power through an identification of technology with masculinity, so that even when women use tractors they are constructed as lacking in technical expertise. Ironically, a tool that makes physically challenging work more accessible to a wider range of people enables the entrenchment of masculine identification with the work (Saugères 2002). This is partly accomplished by a ceaseless iteration of the association between masculinity and technology (Bourdieu 2001; Horowitz 2001; Lohan and Faulkner 2004). As Sam's comment about women operating heavy equipment suggests, for some, a woman's ability to drive a car is almost as notable as her ability to use a piece of mining equipment.

MTR intensifies the technological signification of mining to the point of absurdity. The physical effort of miners has been replaced and overwhelmed by the colossal technology of the equipment and the scale of the job. These giant earthmovers do not simply replace the physical effort of the miners, they reduce it to insignificance. The individual worker is simultaneously dwarfed and magnified by the immense equipment of the mountaintop mine, a contradiction that intensifies the symbolic hypermasculinity of coal mining, sometimes at the expense of the worker.

Part of this hypermasculinity comes from the way people do gender through technology. The equipment and blasting used in mining is culturally constructed as a little boy's delight. In my conversation with Ed and Leann, Ed confessed to Leann, "I'll tell you what. . . . I don't know if I've ever told you this, but, uh, they've used dynamite underground in my mine before."

"No, you didn't tell me that."

"And it is crazy! Because—"

Leann interjected, "You was gonna say awesome." (*Laughter.*)

Similarly, Steve, who works for the DEP on mountaintop sites, told me, "I'm just a guy. I get out there and I play in the mud, and there's equipment, and crashing and banging, you know, it's enjoyable! You know, big explosions going off—you know, there's nothing cooler than a big, rip-snorting explosion."

Ben reported a similar feeling. "I worked [outside] on a dozer on a coal pile. And I had to get them big end loaders and stuff. That's . . . they're intimidating, or whatever you want to call it, 'cause they're huge, too. But once you get up there and get the feel of what you're doing and you feel comfortable . . . it's kind of awesome, you know. You know, you control this enormous thing."

The articulation of coal mining and modernity is gaining new significance as people try to make sense out of how MTR is swiftly and permanently altering the environment. Simultaneously, however, the cultural articulation between masculinity, technology, and a love of risk is a regulatory fiction that, in Butler's terms, can police the expression of emotion about mine work. In Ed and Leann's conversation, Leann interrupted Ed's description of his feelings about the use of dynamite in his mine with the comment that he was going to say "awesome." It is possible that Ed had other feelings about the incident, which he went on to describe: "It don't matter how far away from it you get. When you have an explosion like that in an enclosed area, I mean, you really feel it. . . . I mean, you have . . . to open your mouth. They say open your mouth when . . . it goes . . . that way it hurts your ears less. Because—"

"You could have bust your eardrums with your mouth closed during it," interjected Leann.

"I had my mouth closed in one of them. I . . . couldn't tell the difference between the two."

Leann laughed. "Your eardrums are busted anyway!"

In certain circumstances, it is possible to tell another story. Jesse, who worked on a Massey MTR mine, told me he was "terrified up there" driving a rock truck on the night shift. He spoke to me on the telephone and was not interested in a face-to-face interview. He described the harrowing conditions of a MTR mine. He left for work at 4 p.m., came home at 6 a.m., six days a week, working at a site an hour and a half from his home. For

this, he earned $16.79 an hour. During thunderstorms, he said, the light-
ning is all around you, and the ground on the mine gets slick. Because he
worked hauling rock, he had to back up his truck to the brim of the valley
fill to dump his load, risking tumbling over the side, as a coworker had
recently done. Or the brim could give. He said, "I don't have the nerves for
it. Some guys, it don't bother at all, [but] you're up there intermingled
with the huge trucks." He said of Massey, "Their motto is safety first; it's
really coal first." For instance, the company neglected to replace the fog
lights on his truck, which help him see the uneven ground on the mine,
because, "it ain't foggy." In addition, the truck Jesse drove, a smaller three-
by-seven-foot rock truck, was not equipped with the large rearview cam-
eras that are installed on the enormous nine-by-thirty-foot coal trucks.[9]
Unlike in underground mines, where crews are like "a band of brothers,"
on the surface mine they seemingly don't look out for each other. Jesse
reported that the enormous coal trucks go over thirty miles an hour, and
their drivers aren't concerned with what other workers are doing.

Although Jesse spoke of his deep unhappiness about working on "the
strip," he also voiced what seemed to be the company line about postmining reclamation. He said he worked on a mountaintop mine "the size of
[the city of] Beckley," and it would take a long time to reclaim. But he continued, "Once it's reclaimed, it'll be better. [There will be] small ponds.
The animals like all that. It'll be OK." Jesse had recently had a religious
experience that led him to devote his life to Christ. He spoke regretfully
about the fact that he had taken the job on the strip for the money, thinking this would make his wife happy and save his marriage, but it had failed
to do either.[10] "These strip jobs are a bad thing. I thought I'd find happiness there. All I found was money." This money allowed him to purchase
a new refrigerator, a new living room set, and fashionable clothes for his
kids, but he told me these things were worldly, and that the hours he was
now stuck working to pay for these things were ruining his marriage and
family life. Nonetheless, when I asked him about the spiritual implications
of the notion that the mountaintop mines were improving the land, he
replied, "It would be blasphemy if they said they were going to make this
land better than what God made it. All they are saying is that they are making it more useful."

The idea that humans have the power to change nature in service of
human goals is one of the most striking aspects of pro-MTR discourse. As
noted, a common description of postmining sites, after they have been

"reclaimed," is that they don't look like West Virginia. This (supposedly useful) flat land is frequently mentioned as one of the benefits of MTR. The topography of southern West Virginia is understood as inadequate, or inappropriate, for economic development. This vision of the topography as problematic echoes understandings of the people as problematic, as in culture of poverty explanations that blame coalfield poverty on learned helplessness, or Appalachian fatalism, welfare dependency, or general backwardness. "We don't have any flat land here" becomes the refrain that explains local poverty in terms of local problems, but this time it is the land itself that is at fault and MTR offers the chance to fix it.

This ideology of improvement was evident during a school field trip I accompanied to a mountaintop mine in 2004.[11] The company president acted as tour guide, indicating how important these tours are. Explaining the mine to the kids, he stressed the technical difficulties of MTR. He said that modern coal miners have to be computer literate and that special engineers

Valley fill at an MTR mine, Mingo County, West Virginia, July 2004. Photograph by the author.

had to be hired to carefully calibrate every inch of the valley fills. This claim needs to be understood in the context of other stories circulating in the coalfields about companies "dumping" overburden in valleys and about these valley fills collapsing in deadly floods that are made worse by MTR. Activists and others in the community say that MTR is increasing flooding (Stockman 2004) and is in reality haphazard and careless—they say companies mine as fast as possible and slap together valley fills, spray grass seed on the site, and call that reclamation. The company president's comments are an attempt to refute the environmentalists' perspective with a claim of technical expertise and control. This technical domination of nature represents an ideal of modern masculinity and is part of the everything-is-going-to-be-all-right story that coal companies repeatedly tell local residents.

Some miners and coal industry supporters stressed the difference between what they called prelaw and postlaw. This division refers to the mining practices that went on before the seventies-era environmental regulation.

Valley fill at Twisted Gun Golf Course, Mingo County, West Virginia, 2004. Photograph by the author.

The obvious negative effects of strip mining were instrumental in generating political support for environmental regulations (Montrie 2003). At that time strip mines used a technique referred to as "shoot and shove," meaning to dynamite the earth over the coal seam and shove it over the hillside. Southern West Virginia bears many scars from this kind of mining. MTR operators try to distance their mines and their valley fills from this history by claiming that modern surface mining is an exact science and that they are in total control of its outcomes.

During the mine tour, the company president told us that mining was no longer the low-skill job that it used to be. He described the difficult technical work of MTR, in particular the engineering of valley fills. He commented, "We need to redo the topographic maps every two years," a statement that signals both his assumption of a total domination of nature and the relentless pace of MTR. When asked about the environmentalists' opposition to valley fills, he reassured the kids that they just needed to wait for

School group with excavator at MTR mine, Mingo County, West Virginia, July 2004. Photograph by the author.

*School group in the
dump truck at MTR
mine, Mingo County,
West Virginia, July
2004. Photograph by
the author.*

the finished product, which in many cases is better than the original mountain. "At the end of the day," he said, "it's all about postmining land use."[12]

At the climax of this tour, we got a close look at some of the gigantic
equipment of MTR. The kids got to have a body-level experience of the
power of this equipment by posing for group photos inside an enormous
truck bed. This kind of ritual visit to a coal mine, which is a typical experience for area schoolkids, gives them a special one-time glimpse at the eso-
teric world of colossal mining machinery, which is already legendary. In an
indication of the important symbolic work this kind of visit does, one of
the last things the company president said at the end of the visit was "I hope
you liked the machinery."

The "modern man" story is also present in the paternalistic posture of
the company. After the tour, children were given gift packets with T-shirts
(imprinted with the words "Earth Day" along with the year, company name,
and two white cartoon kids planting trees on a tiny Earth). They also received highlighters, pencils, and pens with the company name, a tree in a
plastic tube to plant, and some informational literature about electricity.
One thing this literature does is put the mine in a context for children—it
is part of the making of the modern world, providing electricity for the

nation, enabling modern lifestyles. West Virginia can thus take part in the American postindustrial economy. The postindustrial nation, we are told, is powered by information and technology; the dirty work of producing things is supposed to be outsourced to the so-called Third World. This modernization narrative naturalizes MTR. By shifting mining away from its association with brute physicality and into a more objectified relationship with nature, MTR moves coal mining masculinity further in the direction of symbolic whiteness, abstraction, and technocratic control. With its articulation with modernity, and by playing on people's hopes of overcoming Appalachian marginalization and poverty, MTR promises a particular future for West Virginia.

After the tour the company provided lunch in the mine office lunchroom, where we listened to a lecture from a representative of the federal government's Mine Safety and Health Administration. He came in and said, "I'm the law around here" and then proceeded to give a lecture about ATV safety. His major points were the dangers of playing around abandoned mine sites, where children might find undetonated explosives or get caught in cave-ins due to subsidence. He also stressed the importance of being careful on ATVs, wearing helmets, and so on. His talk concerned the safety of children who may stray onto a mine site—not the safety of workers, which is presumably his primary focus. The prominence of this government representative during the mine visit did other important cultural work; it identified him closely with the company, which suggests an equivalence between the coal company and the government. His avuncular talk intimated that the company is concerned about children's safety. It gave the mine tour an element of public service, in addition to the large amount of "propaganda," as one teacher put it.

This paternal posture of the company in regard to schoolchildren is echoed in a parallel structural relationship. Because of West Virginia's property tax system, the relation of coal companies to schools becomes a voluntary relation of patronage and publicity (Haas 2008). The coal companies are the major local "partners in education." On the morning of the tour, I waited with a teacher for the school buses to arrive to take us to the mine. She worried that not enough kids would come on the tour. She was nervous because this would be impolite, since the company had donated money for the summer program and had planned a nice lunch for the children after the tour. This is the kind of relationship that gives substance to the phrase "King Coal." Coal company influence is so strong in the area

Partners In Education sign in front of Whitesville Elementary School, Whitesville, West Virginia, 2004. Photograph by the author.

that there is a feudal-like relationship that goes beyond purely economic rationality.

Reworking the Figures: Alternative Stories

The association of masculinity with providership, technological rationality, and domination of nature does not reflect some essential content of male identity. Rather, these articulations are part of a series of gendered discourses in Euro-American culture that give those things coded masculine higher status and greater social power than those things coded feminine, creating a dichotomous categorizational principle or regulatory fiction that involves everyone in a masculine hegemony (Butler 1990; Cohn 1993; West and Zimmerman 1987). The association of masculinity and technology is continually reproduced by repetitions that make it seem natural (Bourdieu 2001; Saugères 2002). The articulations between the status of provider, masculine physicality, modernity, and mining have to be continually re-made and reiterated. They can be disrupted, and the ways that MTR is transforming the masculine content of mining as it alters the daily experience of the work are one example of this. These articulations are also disrupted as anti-MTR activists negotiate these locally hegemonic figures of mining masculinity.

As soon as MTR is placed within a jobs-versus-the-environment framework, it becomes articulated with "masculine" things like providing for a family, technological development, and rationality. Within the terms of the masculine–feminine binary opposition, environmentalism can be read as a primarily aesthetic, unrealistic, and irrational concern. In addition, activism against MTR is characterized by some of the same feminine-coded concerns of other environmental justice movements. For instance, the health and safety of children is central to this movement. The coalfield activists fighting MTR defy the locally hegemonic associations between mining, morality, and masculinity. Activists must "do gender" in the context of hegemonic coalfield masculinity based on physical strength, technical domination of nature, and the provider role.

Because supporters of MTR presume that the men fighting MTR are not miners, they characterize them mostly in terms of failure; they are classed as nonworkers (nonminers), and as such, they are not seen as having much moral standing. This criticism is gendered; women activists are not blamed for not being miners. One of these men, Bill, refused to sell his property to the coal company—they were going to use it for a valley fill. His property sits on a large coal seam, but he stands stubbornly in the way of "progress." He was constructed as the ultimate hillbilly in conversations and newspaper articles that were written about his involvement with a lawsuit. He lived at the head of a "holler," in a "tarpaper shack," local MTR supporters said. Their description was of a stereotypical backward hillbilly; one person laughed and said I'd better take a shotgun with me when I visited him. Another activist, Roger, is in a similar position of standing between the coal companies and a whole lot of coal. His family jointly owns a small piece of property that is entirely surrounded by MTR sites, some currently ongoing, some now reclaimed, and is one of the few places where anyone can come and witness MTR.

Both Roger and Bill are full-time activists. They devote themselves heart and soul to fighting MTR. In contrast to the miners I interviewed, they do not construct themselves as breadwinners; this is not the primary purpose of their work. Instead, they describe their relationships to their families in terms of emotional connection. Each was forthcoming about the state of their marriages, and it was clear that love determined the health of these relationships.[13] Their claim to the position of family men seemed based on being an emotional center for their families, rather than taking the provider role, which shifts the definition of the family from a primarily economic one to a spiritual one, as in the evangelical language of masculine headship and servant leadership (Gallagher 1999; A. Smith 2008). This accords with the largely Christian language and social context of the grassroots anti-MTR movement.[14]

Like most anti-MTR activists, Bill and Roger have been threatened by violence many times. Neither of them easily fulfills the masculine ideal of tough guy, but they both tell stories of standing up to intimidation, especially using logic and rationality to talk their way out of tight spots with people who want to hurt them. This is illustrated by Roger's story of an encounter with some angry mountaintop miners. He was driving down the road from his mountain when three big trucks blocked his way. He stopped his truck and quickly decided on a course of action. He took his gun out and laid it on the dashboard. When the three men approached his window, he made no effort to hide his gun. The men accused him, "You're trying to take our jobs." He got out of his truck and walked up to one of the men. Referring to his relatively small stature he said, "I looked him straight in the belt buckle." He reminded the men about the temporary nature of their jobs and argued that he wasn't trying to steal their jobs; their own friends and coworkers were—and, more important, they were stealing the jobs away from their own sons. He asked them, "Every man wants his son to follow in his footsteps. What is going to be there for your sons?" In the end, the men were no longer angry and went on their way, thinking about what he had said. This version of toughness is someone brave, strong, and articulate enough to stand up to the coal companies and their proxies; people in the anti-MTR movement demonstrate this remarkable quality every day. In Roger's story, the hypermasculinity of coal mining is what enabled him to bridge the gap of understanding between his position, the men's concern for their jobs, and their desire to provide for their families. By linking the provider role to their children's future, Roger was able to catch the angry miners' attention.

Bill's and Roger's retelling of the future promised by MTR reverses the progress narrative of the modern man story and claims a loving, romantic relationship with a feminine nature in place of the "dead" nature that is the object of domination in pro-MTR discourse (Merchant 1980). Roger described the relationship this way: "I told my wife I was having an affair— with this mountain." Another way he described how he felt about the mountains was, "Our mothers give birth to us, of course. But the mountains give us life." MTR is a threat to a feminine nature, in this declensionist version of modernity. Bill also resisted hegemonic instrumentalist understandings of an objectified and abstract nature. He told me that many of the people in his community who had been bought out by the coal company or had lost their homes to MTR had died of heartbreak, and that he could never live anywhere else. His strong connection to place flies in the face of the cultural logic of MTR. Like many West Virginians, he says he was born and raised in his family's homeplace. His extraordinary tenacity in holding out against pressure from the company and its supporters is a refusal of their abstract conception of place. This also allows him to be constructed as a backward hillbilly.

In their arguments against MTR, these men both posit an essential connection between their lives and a particular piece of ground. They use religious references and the emotional power of tradition to defend this piece of ground against MTR. This is a rejection and inversion of the modernist rationale of MTR that claims that the flat land it provides is a benefit to West Virginia as a potential site for economic development. They also both embrace noncommodified, off-the-grid ways of living. For example, they use solar energy panels and well water. Bill sarcastically described the coal companies' version of progress this way: "Mining companies sink creeks, ruin wells, and then put the people on city water and call that economic development! That's improvement!"

Central to the anti-MTR movement is a white woman who became an activist after her grandson came up to her with dead fish from the stream in their backyard, asking, "Grandma, what's wrong with these fish?" She and her family immediately moved away from their homeplace, which had become dangerous, and started fighting MTR. Among the major concerns of the coalfield environmental justice movement are the safety of schoolchildren in a school located near a coal preparation plant, flooding, water quality, toxic sludge disposal, and valley fills. But like Bill and Roger, many coalfield activists have personal stakes in saving a particular mountain, often the site of their family's homeplace. This connection to place might help them to shift the terms of masculine hegemony. Instead of reliance on a job for a family wage, for instance, a strong connection to a piece of land allows a certain level of security and subsistence outside the bounds of the gender-segregated job market.[15]

What is interesting about Bill's and Roger's reworking of these masculinity stories are the ways in which they must negotiate the hegemonic associations in the coalfields between masculinity, mining, and technological and social progress. The environmental justice movement, by fighting for the survival of marginalized people and communities, may promise a relaxation of the heteronormative gender dichotomies that structure many environmentally harmful practices. This is apparent in the centering of feminine-coded concerns in environmental justice activism (Stein 2004, 5). Without any overt challenge to sexism or heteronormativity, the coalfield environmental justice movement presents an alternative interpretation

"In His hand are the deep places of the Earth. The strength of the hills is His also": a memorial for the mountains marks the entrance to a family cemetery surrounded by mountaintop mines. Raleigh County, West Virginia, August 2004. Photograph by the author.

of some of the most hegemonic signs of American heteropatriarchy: technology, progress, and the jobs-versus-the-environment aphorism.

Clearly, it is the story of modern man that is the most challenged by anti-MTR activism. This story is central to creating the cultural context that makes MTR make sense. These figurations of mining masculinity are especially charismatic in the local context, where they evoke the national symbolic and refute the backward hillbilly image. But as the coalfield environmental justice movement has grown, it has launched an increasingly powerful critique of the deeply entrenched white heteropatriarchal terms of this coal-mining logic, bringing the fight against MTR together with other movements against the exploitation of nature and marginalized communities.

White masculinity, especially its incarnations as family man, tough guy, and modern man, makes up part of the cultural logic of MTR. This cultural construction cannot be separated from the material conditions in which MTR exists. The gender- and race-segregated job market works together with cultural institutions such as heteronormativity to create the conditions of possibility for unthinkable environmental destruction. But the articulations that make up this cultural field are not essential or fixed. They must be re-created and reproduced constantly, and they can be shifted and rearticulated in new ways, creating new political possibilities. In any case, resistance to environmentally devastating practices like MTR is made possible in part through a material and cultural reworking of the hegemonic identity formations that support these practices.

CHAPTER 3

The Gendered Politics of Pro–Mountaintop Removal Discourse

THE TINY TOWN OF BLAIR, West Virginia, has a violent history as the site of a famous battle in the Mine Wars of the 1920s, when militant union miners fought to install the union in Logan County by force (Savage 1990; Shogan 2004). Nearly eighty years later, Blair was the site of another conflict, this time over MTR coal mining. In 1998 residents filed a law-- suit against the West Virginia Department of Environmental Protection (DEP) for failing to enforce regulations that call for a buffer zone to protect streams. Although surface mining regulations theoretically require com- panies to return the rocks and soil of the mountain to their "approximate original contour," in practice the DEP typically grants exceptions to this rule, and mining companies almost always leave their mining sites flattened and surrounded by valley fills. Because streams are protected by the Clean Water Act, valley fills have given environmentalists a legal front for their fight against MTR.[1]

At 3,196 acres, Arch Coal's Daltex mine in Blair was the largest MTR mine ever permitted in West Virginia (Ward 1998). Before the lawsuit, it had already permanently altered the topography around Blair and effectively destroyed the community. While the mine was working, the company bought up houses in Blair in order to prevent lawsuits over damages and to be able to use the land in the mining operation; it took back the land it had donated to the state for a school in order to use it for a coal preparation plant. As the mining progressed, the community of Blair was eroded, as more and more of the residents were bought out and moved away. Sarah, a white resi- dent of Blair and spouse of a retired miner, described for me the effects of the working mine. Coal dust and dirt filled the air, making it difficult for her hus- band, a former miner with black lung, to breathe. The blasting nearly shook her out of bed one night. Her house was damaged. Her well was destroyed.

When Judge Charles Haden called for enforcement of the buffer zone rule, Arch Coal shut down the Daltex mine, laying off hundreds of workers and dividing the community (Ward 1998). With the mine silenced, the community continued to decline. The town was dismantled; the schools and restaurants were gone. For a few of the current and former residents of Blair, it was difficult to say which was worse, the hazardous conditions caused by the twenty-four-hour blasting and shoveling of an ongoing MTR mine or the economic death they related to the mine closure. Although MTR made the town uninhabitable, coal industry supporters blame the mine closure for the deterioration of Blair. The lawsuit made enemies of community members and brought former friends to blows.

MTR is simultaneously destroying forests and ecosystems, flattening the beautiful Appalachian Mountains, fundamentally redefining the job of coal mining, and reducing demand for workers through increasing mechanization, yet the defenses of the industry offered by many coalfield residents often do not distinguish about how the coal is mined. The industry works hard to promote the image that "Coal's been good for West Virginia," using public relations strategies that play on affective and charismatic images of masculinity and local pride. In 2008 the West Virginia Coal Association ran a billboard campaign featuring local heroes like Don Nehlen, the former coach of the West Virginia University football team; Bob Pruett, Marshall University's former football coach; and professional bass fisherman Jeremy Starks as "Friends of Coal."

This chapter focuses on how the mutually productive intersections of race, gender, and class shape political culture as well as perceptions of environmental and economic risk. The love of mining, and of MTR, is related to a particular definition of freedom that lies at the heart of U.S. political culture, in ideals of citizenship and belonging to the American community. A gendered understanding of work, embodied in the heterosexual white male breadwinner, gives shape to a specific configuration of masculinity that gains moral worth from family-wage employment. I label this formation of masculinity "dependent" in order to problematize dominant constructions of independence and dependence, public and private, that reproduce gender inequality. Through its articulation with this form of masculinity, mining achieves more than simply economic importance in the coalfields; it also maintains proper American citizenship.

Masculinities are important not only to the men who embody them but also to the rest of their communities because they help structure complex social relationships (Connell 1995). Masculine hegemony, or the way that masculinities dominate the "real," is structured by a multiplicity of masculinities and femininities framed in relation to one another. Dependent masculinity works to maintain an ideology of masculine economic primacy but is itself dependent on a privileged but vulnerable position within a gender- and race-segregated job market (Fine et al. 1997). The relationship of this form of masculinity to other hegemonic masculinities can be seen at work in the perspective of white-collar workers who represent a dominant American ideology of economic rationality and independent citizenship.

Anglo-American narratives of individualism and a naturalized, prepolitical market mask this positional dependency and give it its normative content (Somers 1995). Although formal citizenship is conferred by birth, ideal American citizenship is related to meeting certain moral standards of self-sufficiency. This political culture simultaneously problematizes working-class citizenship and interpellates the miners and the coal companies as rational individuals operating in the free market. American national identity relies on this ideal of rational, independent individuals, the sort of individuals who are understood to be capable of self-governance. Once seen as a quality limited to white male property owners, this condition of independence has continued to be problematic for women and people of color (Goldberg 2002, 48; Mink 1990). Historically, the white male working class was also one of these problematic groups, but the moral status of breadwinner offers a resolution to the problem of independent citizenship for white working-class men. This American story of morality and freedom helps explain why some people tolerate MTR.

Citizenship and Masculinity

Support for MTR arises in the context of certain values: a particular definition of freedom, a masculinity of providership, and moral citizenship. The importance of white breadwinning masculinity to ideals of American citizenship makes a mine closure a morally threatening event and a danger to a particular kind of community. As long ago as 1990, retail made up almost as large a percentage of total employment in Boone, Logan, and McDowell Counties as coal mining, and in 2000 Walmart was the fourth-largest employer in Logan County (Burton et al. 2000), but this type of low-wage employment was not discussed in my interviews.[2] When people talked about work, they were talking about men, particularly family men—breadwinners. Typically, a job at Walmart is understood not to be adequate to take care of a family. In southern West Virginia, the kind of work that can support a family has been synonymous with coal mining for much of the twentieth century. The gendered structure of work, class, and race divisions in neighborhoods; types of housing construction; and community planning in general all evidence the dominance of coal (Barry 2001). Place-names like Charmco or Holden, derived from company names, reflect this historical pattern. Today this spatial dominance continues, as coal companies relocate the residents of former coal camps into new trailer parks to make way for new operations, at the same time as upscale housing developments are built on reclaimed surface mines.

Mining is considered the driving force behind the economy in southern West Virginia (Burton et al. 2000). For much of the twentieth century, this has been the case. People often remark, "If it weren't for the coal industry, we wouldn't be here." Today, employment in coal mining has decreased sharply, and communities are no longer centered on one company or one mine as they were in the past. Even if the companies the towns are named after are gone, a job in mining is still a unique and important opportunity for local working-class people.[3] The fact that coal-mining wages are so much higher than wages in other types of employment, especially in more feminine-coded fields, is not an accident of nature. It is a result of the very family-wage ideology that is now being used to support MTR.

Nancy and her grandson Greg are former residents of Blair. Nancy's late husband was a miner, killed in a mining accident as a young man. Greg is a young miner who left Blair when the mine was shut down. Nancy had moved away earlier, in part because of the dangers and inconveniences caused by the blasting when the mine was running but also due to her age; she is elderly, and her children wanted her to move into a larger town. Although Nancy missed the house where she raised her children, her attitude toward the mine closure was mixed. She seemed to regret the mining jobs that were lost to the men of the community as much as she regretted the personal loss of her home. Here, she clearly articulated the gendered definition of work, which equates work with masculinity. At the same time, she located the personal sacrifice of her home within a community morality of masculine work.

> I think they should have stopped it to start with, when they first
> started destroying everything and they saw what they were doing
> and we could have continued to live there. Or if they were going
> to make us move, they should have continued to work. They'd

already destroyed the place, why didn't they let the men keep working it? People have got to work. . . . I didn't like having to leave, but after I had to leave, and we were all torn up, I think we should have let the mine work. . . . Blair could have been saved. But after they destroyed our places to live, the men should have been able to work.

For Nancy, the need for men to work in mining seems to outweigh even her own attachment to the physical community of houses and people. Her comments reveal a tension between two competing ideas of community: one the concrete locality of houses and people, the other an abstract moral community of nuclear-family households. The environmental and economic impact of MTR makes clear that these two ideas of community are incompatible.

Discussing how many people have been affected by the mine closure, Greg posited a family of dependents for each worker, who is assumed to be a family man and a breadwinner. His argument reveals the moral content of coal-mining work as essential to a coherent community of heterosexual households:

> You can't just say that four hundred men lost their jobs. . . . Let's say for every one person who works at the mine there's ten support jobs, so you take four hundred times ten and you're looking at four thousand people that'll be affected. And then there's three people you can say in the household, then you multiply that by three and that's twelve thousand people, just by one little job.

This definition of the male worker as the head of the household structures what counts as real work and therefore conceptually limits the economic choices available to the community. It also places the male worker in a complex relationship to the dependent members of his household, the state, and the market. Greg represents these jobs in mining as the lifeblood of a particular community, a moral community of heterosexual male breadwinners and their dependents, which is not necessarily tied to a physical community and, indeed, in global capitalism is perhaps incompatible with the existence of a physical community. This moral community of heterosexual households is also invoked in Nancy's comment about the environmentalists who supported the lawsuit: "They come in, they get all upset, and they throw our men out of work."

The centrality of work and the identity of breadwinner in masculine subjectivity and community identity is located in a genealogy of industrialization and racialization. Masculinities are historically variable and multiple (Connell 1995). Theorizing multiple masculinities as part of a historically specific configuration of gender, race, and class formations is key to understanding how the gender logic of MTR works. Although today it seems natural that masculinity is "tied up with the breadwinner role," in other contexts the structurally dependent position of working-class men and other laborers on their employers placed them in direct contradiction to the hegemonic ideal of (masculine) citizenship (Fraser 1995, 33).

In the early "republican" phase of U.S. political history, the success of democracy was considered to be linked to independence. Only an independent citizenry would be capable of rational judgment. At the time, true independence meant owning property. People of color who were expropriated or enslaved in this system clearly did not qualify for the masculine state of virtuous citizenship, and neither did the white propertyless class. With the institution of white manhood suffrage, race and gender became more strongly determinate than class in defining citizenship. According to the gender ideology of republicanism, women inhabited a nonpolitical private sphere. Masculinity consisted of independence and interaction in the public sphere and was opposed to a feminine condition of dependence. This feminine condition of dependence was also ascribed to ethnic men and men of color, who were prevented from taking part in the public sphere. In American political culture, independence and citizenship thus became articulated with masculinity and whiteness (Hurtado 1996; Mink 1990).

White manhood suffrage gave even propertyless white men full citizenship. Within the terms of liberal democratic citizenship, this redefined white men as "independent," self-reliant political actors against "dependent" white women and people of color. This was managed by the elevation of the white wage laborer, who had previously been considered a dependent of his employer, to the status of an independent head of household who (ideally) earned a family wage. Hegemonic masculinity was transformed in this new historical context to include the white working-class breadwinner. Propertyless white men could now be counted on to uphold the national ideal. Whereas freedom and independence (citizenship, in short) had formerly been based on property ownership, they could now also stem from the ability to sell one's labor—specifically the privileged labor of the white male working class that could potentially earn a family wage. Labor market segmentation and a gender-based wage system solidified the privileged position of white male labor (Hartmann 1976; Roediger 1991).

As industrialization reorganized space and segregated cities by industry, the job became a greater signifier of both masculine identity and community identity (Hoggart and Williams 1960). Working-class communities are still typically identified according to what the men do: a place is defined as a steel town, a mining town (or, more likely, a former one). The non-coal-related jobs available in southern West Virginia, regardless of wage, do not carry the same cultural weight. In particular, female-coded types of employment do not typically identify communities. The family wage ideology tends to erase the service-sector work often done by women at the same time as it justifies the low pay earned by these workers (L. Gordon 1994). These gender ideologies and moralities around work impede the mobilization of coal miners and their communities for an environmentalist cause.

According to the logic of the family wage, the significance of work varies greatly according to whether men or women are doing it (Flores 1997; Mohanty 1997). When a high school graduate takes an entry-level job in a coal mine, he is not simply making an economic decision based on wages and rational choice; he is taking a moral place in the community as a provider. Ideal white working-class masculinity hinges on men's ability to fulfill the role of breadwinner and therefore be independent and moral citizens. Ironically, the "independence" of this white working-class breadwinner, which is constructed against the "dependence" of others, is itself utterly dependent on the availability of high-paying jobs that have been generated by the inequalities of the wage structure and the labor market. Although hegemonic in the sense that it contributes to the logic of the status quo, white working-class masculine identity is marginalized in relation to its dependency on the decisions of managers, CEOs, politicians, and policy makers. In the formation of dependent masculinity, the structures of heteropatriarchy and the capitalist economy work as one. Its existence is dependent on the very structuring of the labor market that has generated its unique and defining privilege of earning a family wage. For white working-class people virtuous citizenship depends on participation in a gender- and racially segmented labor market that privileges white working-class men over racialized men and all women.

Accordingly, for many of the white coalfield residents I interviewed, working at a job with a family wage seems to have a great moral significance

ımunity. The extent of this significance is revealed in the appar-
ness of men and their families not only to sacrifice the environment but also to risk bodily mutilation or death to work in the mines.
Among the working-class people I interviewed, those who were working
themselves or who were related to workers tended to distance themselves
morally from the category of nonworkers—again, with "work" being closely
linked to mining. The proper masculine subject is a worker and a provider,
and not working casts doubt on a man's moral character (and by extension, that of his family). In Blair, the local environmentalists who filed the
lawsuit were criticized by others for not being workers. Nancy dismissed
their opinions this way: "There's three or four in Blair that have backed
these environmentalists and they never did work anyhow! They don't care!"
Regarding one local environmentalist, she commented, "Let me tell you,
this man never worked. I don't know that he ever worked in the mines." In
her dismissal of their point of view, Nancy indicated that because the environmentalists didn't work in the mines, their interventions were invalid.
At the same time, she isolated these "three or four" people who opposed
the mining and associated herself with the community of workers. This
reaffirms the morality of work and underlines the articulation of "working,"
morality, and supporting MTR.

Alicia, another resident of Blair and the wife of a retired miner, used
similar words to describe another environmentalist from Blair. She commented that at his house, "The grass is up to your knees, and [there are]
pop bottles and cans out in the yard . . . [the house] is just a shack." I asked
if he was a miner. She responded, "Well, I don't know of him ever working." In Nancy and Alicia's comments, "working" is related to support for
MTR (or at least not vocal opposition), and those locals who oppose it are
characterized as nonworkers, which undermines their moral status. Similarly, Richard, a white county official, called one of these men "the trash of
the earth." As Nancy and Alicia's remarks illustrate, women as well as men
are invested in this morality. It seems that this type of masculine identity
is important to the imagined community as a whole (B. Anderson 1983).

A middle-class perspective clarifies the stigma of "not working" and its
relationship to the culture of dependency often attributed by outsiders to
working-class people and poor Appalachian communities. Chris, a white
mining engineer in his forties, explained West Virginia's problems in terms
of this perceived local culture of dependency. In his view, people's passivity and unwarranted sense of entitlement to government handouts were

largely responsible for coalfield poverty. This poverty is exemplified by the lack of a public water system in McDowell County. Chris mentioned wells with questionable water quality and the lack of a public sewage system outside the county seat. When I noted that this degree of poverty is remarkable, considering that McDowell County was once known as the "Billion-Dollar Coalfield," Chris responded in terms of the culture of poverty:

A lot of it is passed down, from parent to child; they don't know any better. And I grew up in Mingo County [another coal-producing county in southern West Virginia] and I understand it; it's a very difficult bond to break, that you knew that Mom and Dad went to the mailbox the third of every month and got a check. They never worked, and you, you do the same thing.

From his perspective as a middle-class, white-collar worker, Chris views coal communities "objectively," clearly separating himself from the "problem," and blaming the extremely poor for their own poverty. Despite the well-known fact that the coal industry took untold wealth from the area, using the labor of the residents, and left when the coal was gone, these comments relate the lack of public infrastructure to the moral worth of the people. Because there is no more work, the people are nonworkers. Unlike Nancy, who emphasized the difference between the community of independent working men and the nonworkers who opposed the mine in Blair, Chris's comments indicate a suspicion that all working-class coalfield residents are predisposed to dependency on government assistance and hence are not proper citizens.

Despite his reference to "Mom and Dad," Chris and his friend Rob, another white engineering firm employee in his forties, only discussed jobs in traditionally masculine fields. They also limited their comments to working-class jobs; in this conversation, coalfield residents were always already manual workers. This move further identifies the problem of achieving independent citizenship with a working-class Other, not with the middle-class self. When asked what other kind of work was available in these depressed coal communities, they listed masculine-coded jobs: "a rebuild shop, they rebuild motors . . . the local Caterpillar dealer, that sells equipment," in addition to spin-offs of mining like cement plants. In this view, the possibility for coalfield residents to achieve a moral state of independence, in opposition to dependency on government aid, comes from distinctively masculine types of manual labor.

But if Boone County's largest industry is coal mining and its spin-offs, in the surrounding coalfield counties of Kanawha, Logan, and McDowell, services, government, and retail rank higher (Burton et al. 2000). Notably, in our discussion of available jobs outside mining, Chris and Rob did not mention the service-sector jobs at the Charleston area's new mall or at the Logan Walmart. As noted, by 1990 almost as many people worked in retail as in mining. But the wages for retail have never compared to "miners' pay." In fact, good coal-related and spin-off jobs are few, and the traditionally high pay is not guaranteed in today's business climate of capital-intensive mechanization and union busting. But they remain the only type of work imaginable for a certain type of person—the independent head of household. Other types of work, low paying and female-coded, are not even worth mentioning.

In these comments, the power of the family-wage ideology is clear. Work is defined along gendered lines and related to morality and citizenship in complicated ways. The definition of propertied independent republican citizenship from the early United States has been reconfigured in the story of the family wage. Its logic doesn't need to conform to empirical reality because its power as a narrative comes from its internal logic (Somers 1995). The masculine independence of the head of household is constructed in opposition to the perceived dependence of his wife and children.

Masculinity and U.S. Political Culture/Nature

Masculinity helps shape the relationship of the miners to the coal industry. But the narratives of individualism and the naturalized market that characterize U.S. political culture mask the dependency of the workingman on his employer. This political culture discounts the state's ability to function effectively at the same time as it muddies the question of exactly what kind of competition is occurring and between whom. For the people I spoke to who supported MTR, federal and state regulation were perceived as interference in the natural functioning of the market. For some, the union constituted an unnatural interference as well.

Chris and Rob told a story that illustrates how the union is blamed for its interference in the free market. A coal company president offered hundred-dollar savings bonds to schoolchildren with perfect attendance in Mingo

County. In response to this gesture of goodwill, Chris said exasperatedly, "The locals went ballistic." He sarcastically explained that people were angry because the company president "was trying to teach . . . their kids, that it was more important to go to work sick than . . . to take . . . their rightfully allotted time off being sick." Here, Chris portrays the legal rights of workers, specifically rights attained through labor laws and other regulatory measures, as a source of dependency for working people. In Chris's version of events, the people rebuffed the company president's attempt to teach their children values meant to lead them out of their condition of dependency; a true work ethic apparently requires workers to be willing to go to work sick.[4] Elaborating on this perceived dependency, Rob offered the following explanations for what he described as the entitlement culture in southern West Virginia: "The strength of the union, and the way people have come to depend on the federal government to make a living, by welfare, or unemployment, or SSI. All these items come into play. [But] the coal companies—there's not a coal company job in southern West Virginia that would pay less than $50,000 a year."

Here, Rob contrasted the (increasingly rare) family-wage jobs offered by the coal companies with the culture of dependency resulting from government entitlements. In Chris's story, the company president's attempt to teach children proper values, perhaps to form them into suitable future employees, was rejected by the community in favor of their alleged dependency on entitlements. Interestingly, in his analysis, Rob placed the more commonly accepted federal assistance programs of unemployment insurance and supplemental security income (SSI) in the same category with the frequently stigmatized "welfare." In light of the dual nature of the U.S. social welfare system, which is shaped by family-wage ideology (L. Gordon 1994), Rob's comments suggest that he sees even these masculine-coded programs for unemployed, aged, or disabled working-class people as a threat to masculine independence. These stories oppose the morality of responsibility and independence (regular attendance at school and work) to an immoral state of dependency attributed to poor people perceived as living off of the union-inspired social programs of federal government.

Some of Rob's feelings about federal assistance may be related to the racialization of dependence in the United States. In the 1980s, the Reagan administration led the conservative backlash against the War on Poverty of the 1960s, coding welfare dependency as black and female in an oblique national discussion of race (Gray 1995). This feminization and racialization of the social safety net opposes those who rely on it to the good, taxpaying (white) citizen interpellated by the conservative backlash (Hall 1988). In Rob's comments above, the union is identified with the government as an agency that interferes with individual independence.

In the mid-twentieth century, the United Mine Workers of America was able to influence both local conditions and federal policy, and this legacy complicates local understandings of working-class power. Laura, the educator and daughter of a union coal miner introduced in chapter 2, told me that when she was a child her father would sit silently in union meetings only to come home and "let off steam" about the excesses of the union leaders. In her interpretation of her father's opinion, it was the activists in the union who insisted on striking that made her family poor. Her father was a faithful union member, but the frequent strikes frustrated his desire to earn a living. Her comments reflect a view of the union that sees its value as a fraternity of workers instead of as a political bloc. This apolitical view of the union also locates the union along with the company and the government as one of the many powers that working people must contend with in their daily lives. Hence, ironically, it is possible to understand the very achievements of the union in guaranteeing pensions and health care as a *loss* of dignity for the workers, as allocations that reduce the independence of their intended beneficiaries.[5] Two current Massey Energy Company employees that I spoke to also voiced this perspective, stating that the only reason a miner needs a union is because he doesn't want to work.[6]

The independence that is essential to citizenship is a point of convergence between the U.S. version of political democracy and free-market capitalism; the democratic citizen is presumed to be an active participant in the market. The MTR controversy shows how dominant political discourses of individualized competition and a naturalized market effectively disguise the significant dependency of the masculinity embodied in the breadwinner/worker. The ideologies of individualism and unregulated competition are essential in the construction of a free, white-coded, American subjectivity, and the construction of public and private spheres helps create and sustain a gendered definition of the worker and citizen as masculine. This Anglo-American version of citizenship is based on a story about the original state of nature, a world of independent, rational individuals who came together to form the state through a social contract. This theory/story naturalizes social forms like the market economy and the current form of representative democracy (Somers 1995).

In her discussion of the metanarrative of Anglo-American citizenship theory, Somers includes a system of binary oppositions along the familiar lines of public and private spheres. This system of binaries contrasts in interesting ways with the feminist analysis of the Western culture/nature binary, which places culture in the privileged category (Ortner 1972). Although some have theorized that the culture/nature binary is a universal foundation of patriarchy, that claim has been solidly critiqued (Alarçon 1997; Carby 1982). I am considering it here strictly as a characteristic of Western modernity and liberal political theory (Latour 1993; Pateman 1989; Said 1978).

Anglo-American citizenship theory places rationality, the private sphere, and nature together in the dominant position, which seems to contradict the Western dualisms of culture/nature, male/female, etc. The discrepancies between these two binaries delineate the space inhabited by the masculine subject of society. In these two schemas, words like *nature* and *public* are floating signifiers; they are empty signs whose meaning cannot be fixed.

Anglo-American Citizenship Theory

society	the state
private sphere	public sphere
reason	arbitrariness
nature	culture
science	tradition
free market	regulation

Culture/Nature

culture	nature
achievement	ascription
politics	personal
reason	emotion
freedom	subjection
masculinity	femininity
public sphere	private sphere

Two binaries in Western modernity: Anglo-American citizenship theory (Somers 1995) and culture versus nature (Pateman 1989). The first maps a space of natural political freedom, and the second charts the space of human subjectivity. Confluences and discrepancies in these binaries delimit the uncertain terrain of working-class masculinity.

rather, their meaning is relational and context dependent (Hall 1996). For example, the meaning of nature is different in each set of binaries. The Western culture/nature dualism reflects a definition of nature that includes reproduction and "objective" nature (the environment). In this set of binaries, human agency is opposed to and valued over natural processes. However, in the social naturalism of Anglo-American citizenship theory, nature means something else entirely. Here the natural is made up of things like the market, free associations, personal identity, and objective knowledge, which are thought to preexist the social contract. And while the culture/nature binaries assign overtly gendered characteristics to culture and nature, the binaries of Anglo-American citizenship theory are seemingly genderless. Far from being ungendered, however, this theory describes a distinctively white, masculine subject, as Mink shows in her analysis of the "independent" citizen (1990).

In Anglo-American citizenship theory the natural is the realm of the prepolitical collection of independent individuals that make up society, a realm of natural freedom, limited only by the despotic actions of the sovereign or by arbitrary regulations issued by state bureaucracy. But in the culture/nature binary, the natural is that which is nonhuman, i.e., Other. The masculine subject of society/culture occupies a space of potential freedom in both of these schemas. On one hand, the masculine subject of society enjoys a prepolitical natural freedom, which is opposed to the artificial limitations of law (Somers 1995). On the other hand, the masculine subject of culture is in a condition of freedom opposed to objective nature, which includes his dependents and culturally exploitable natural resources.

These schemas map the uncertain terrain that the working-class masculine subject must navigate between unnatural interference from the state, threatening his fragile moral status as an independent individual, and a feminized state of dependency resulting from the loss of (or failure to gain) this moral status. Coal mining is a fruitful site for exploring these contradictions because mining has a long history of abject working conditions and hyperexploitation of its workers. As Chris's and Rob's comments illustrated, coal miners' claim to the normative morality of independent heterosexual households is quite fragile, but due to the interworkings of these two schemas, it is simultaneously possible to see miners as excessively masculine. This signals the racial valences of these cultural forms. Coal miners are often symbolically associated with the color black. For instance when miners came home from a shift in the (underground) mines, "all you could see was the whites of their eyes," Paul, a miner's son recalled, because they were covered in black coal dust. In a move similar to how African American masculinity is racialized, miners are seen as excessively physical, too close to nature. Paul recounted how his father frightened him and his brothers because he never talked; instead, he "growled." The incredible power of the mining workforce—unearthing millions of tons of coal each shift, providing energy for the nation—is culturally ambivalent, suggesting both strength and an animal closeness to nature.

As such, the whiteness of the white coal miners is also problematic. Again, the othering of coal miners parallels the othering of hillbillies. Appalachian whiteness is an example of a marked whiteness, "whiteness plus something less" (Frankenberg 1993, 198). Hillbillies are seen as too close to nature, too traditional, and too lacking in modern middle-class consumption patterns—they are white people who fail to be white enough. Coal miners are part of the so-called labor aristocracy, that privileged, mostly white working class that benefited from the segmented labor market in order to earn a family wage. But coal miners simultaneously represent a failure of the white labor aristocracy to eliminate the shame of abject laboring bodies (Cohen 1997). The construction of the liberal democratic citizen is related to the privileging of the abstract reason over the limitations and needs of the body (Kazanjian 2003, 19). Historically, however, coal miners' bodies have been literally used up in the course of a life in the mines; backs permanently bent from working narrow seams, lungs permanently choked with coal dust. As Luke, a white underground miner, put it, "By the time I'm old enough to retire, I'm gonna be broken down." The miner's body is mixed with the coal.

The historical habitus of mining life, both in the underground mines and in the coal camps, eludes any attempt to deny the abjection of the capitalist mode of production. At home, coal camp houses were cheaply constructed and covered with coal dust. At work, the coal miner's body was subjected to extreme discomfort, danger, darkness, and exhaustion. The space of freedom constructed for the working-class masculine subject between the two binaries is revealed to be illusory. Because he is neither safely independent from the protection of state regulations nor separated enough from nature, his moral status is constantly in question and must continually be rewon.

In the interworking schemas of Anglo-American citizenship theory and the nature/culture binary, whatever is defined as natural is considered objective and universal. Categorizing things this way removes them from the realm of the political, or that which has to do with the allocation of power and resources (DuPuis 2002). Each schema limns the political into a truncated public sphere. Both schemas work within modern capitalism in the United States to define and delimit what is political and what possible alliances can be made. According to Anglo-American citizenship theory, the so-called free market is part of a prepolitical natural society. Belief in the legitimacy of the market prevents coal miners from questioning the companies' decisions. The companies portray themselves and are portrayed as rational, independent private actors, competing with other individuals in a rational, natural system.

Greg told a story in which his father educates a car salesman about MTR and its role in West Virginia. The car salesman was against the practice, he said, because he had read about it in the paper, the *Charleston Gazette*. Greg starts by quoting his father:

> "Look, I'm buying a $30,000 vehicle. Where do you think this money comes from? I work at a strip mine. Sit and think how many vehicles you sell to coal-related jobs, not just to people who work directly for the mines but all your hydraulic shops and machine shops. . . . All those people that's buying vehicles." Well then, you know, after he kind of talked to them, and they started thinking about it, well, you know, it may not be that bad. When you start cutting into their money—say you sell ten cars this week. Everybody shuts down and you don't sell one, maybe one or two. That wouldn't have been that good. . . . I believe a lot of the people, they really, they have no idea what the whole situation is, they've read stuff about it, read negative [things]—it's bad, it's wrong, it's killing the environment—that's what they're basing it on.

In this story, individuals pursuing their private interests are the motors of society, just as coal mining is the multiplier in the economy. Notice that the individuals in question here are seemingly unmarked yet totally identified with their potential to earn a family wage. "Sit and think how many vehicles you sell *to coal-related jobs.*" The market is a natural meeting place for free individuals with something to sell, whether it be labor power, coal, or cars. State regulations are arbitrary and contingent; in fact they hurt more than they help. This forecloses any questioning of the coal companies' pursuit of profit that drives their push for easier regulations of MTR.

In the naturalized market, the superintensive technique of MTR is an outcome of competition and the drive for efficiency. These concepts, which are fundamental elements in the functioning of the market, are not available for questioning. Greg commented: "A lot of people say, well, they ought to go back to deep mining and forget about strip mining. Well, West Virginia can't just do that and all the other states do the MTR and produce coal . . . at a dollar a ton and here it takes five dollars a ton, well, you can't sell it competitive with them." Mining engineers Chris and Rob also mentioned the competition from the enormous western area mines. Interestingly, this parallels the arguments of environmentalists who point out that the same companies operate mines in the west and the east simultaneously, which nullifies the notion of competition between companies and emphasizes the competition among states to *attract* mining companies. The nuances of exactly who is competing for what are erased, however, in the discourse of the naturalized market.

For defenders of MTR, the companies and workers are understood as individuals bringing their particular strengths to exchange in the market. The companies are portrayed by their supporters as hyperrational actors, perhaps even as ideal individuals. Chris characterizes the companies' position like this:

> We don't need to run these coal companies off. . . . They're willing to do whatever it takes to keep mining coal. So let them build roads, let them, you know, put in water lines, let them do things like that, because until then we're not going to have any other infrastructure. . . . I mean, nobody's going to go in Buffalo Creek in Logan County and build a warehouse, with the nearest water line being seventeen miles away. But people are willing to say, "Yeah, we'll donate money, we'll clean up the streams, we'll do that. Just allow us to continue making this money, our stockholders making this money, off of mining coal."

Here the people of West Virginia are potentially rational, formally equal participants with the companies in the functioning of the market. They only need to work with the companies and not against them. According to this argument, the only things holding back the companies from generously sharing their profits with the community are the burdensome regulations that limit those profits.[7]

Anglo-American citizenship theory places a premium on rational, self-interested, individual agents pursuing their interests in the market. This prevailing definition of citizenship, in which voting is the axiomatic political activity, places miners on formally equal footing with the "individual" corporation and makes collective political action difficult to imagine. For the individual subjects of this citizenship theory, freedom means freedom from the interference of the state.

The market is naturalized and privileged over state intervention and regulation. In Chris's view, the federal government is unable to affect change in poverty-stricken areas of West Virginia; government programs only lead to dependency. Instead of the government, the coal companies are seen as the real power and source of potential economic development. According to Chris, "I mean the money is not here—the only way [to get the infrastructure] is to let these big companies mine the coal now and force them to reinvest in West Virginia, which most of them are willing to do right now." I asked for examples. He continued, "Well, how would they like to do it, is if they're, I mean, they're willing to invest in schools, in transportation projects, you know, building roads . . ."[8] Here Chris credited the industry with more power and ability in this arena than the state or federal government. Since 2000, a number of development initiatives have been undertaken in southern West Virginia, frequently on the flattops left by MTR; these include a golf course (owned and operated by the mining companies), a regional jail, and a veneer factory. In the coalfields, where the coal industry has suffocated other economic activity, the naturalized market forces that justify MTR are seen as the most effective way to build public infrastructure. The government's feeble interventions are seen as not only ineffective but actually harmful inasmuch as they create dependency.

The Way It Is?

Even those who had a less enthusiastic view of the coal industry, such as Nancy and her neighbor Jack, a retired miner, nonetheless saw the functioning of the companies in the market as inevitable. Nancy connected the condition of the community intimately with the coal industry's lot. Describing the poor condition of several area towns, she commented, "Madison has gone down . . . at one time Madison downtown had theaters and everything . . . but when the coal industry goes down, and the work goes down, everything goes down." (Here, her comments imply a link between the economic health of the industry and that of the community that is not borne out by coal production figures and industry profits.) Nancy told me she loved animals, but she couldn't allow nature and the environment to be her top priority. She, Greg, and Jack emphasized the importance of jobs. Greg argued, "What's going to better from shutting [the Daltex mine] down? From that area, what's going to better? In my opinion, everybody will come out better if they mine it, because you're going to have forty-five hundred jobs there—most places would do anything if you'd get that many jobs into their area."

In his deep, gravelly voice, Jack asserted: "I say mow them all down—just look like a flat land all through this country—for a job, wouldn't you?" And Nancy echoed, "Amen! People have got to eat! If it kills a little bird, and I love birds more than anybody..." Jack added, "Kill a bird or a fish..." And Nancy said, "I'd rather see my family eat. That's what it comes down to. Now that we've sacrificed for it, we should have a job." Greg agreed, "Most would say so."

This conversation underlines the importance of a job, and a particular kind of job, to coalfield communities. The importance of a job overrides both the environmental and social devastation caused by MTR. Once again, "a job" is equivalent to mining, and no mention was made of jobs in the service sector. However, the bald assertion that if we want to eat we have to mine is not that easily defensible. There are other ways to get by, through service-sector employment, for one example. This view also ignores the bleak economic reality of coal mining in West Virginia; as has been noted, more coal is being taken than ever but with fewer workers than ever. Even more strikingly, this argument discounts the true environmental cost of MTR, which alters the water table and increases flooding, endangers communities, destroys forests, buries the headwaters of rivers, and forever changes the topography of the Appalachian Mountains.[9] Coalfield communities are receiving an ever-smaller portion of coal industry profits while suffering more and more environmental damage. But if the claim for the necessity of mining doesn't make sense in economic or environmental terms, it makes more sense in the context of a kind of masculinity that relies on the existence of a family-wage job for the status of an independent and moral American citizen. The fact that the independence and morality of the male head of household encompasses the community in a wider sense is indicated in Nancy's "we."

People's explanations for their support for MTR reveal connections between local and regional particularities and the national economic struc-ture, political culture, and hegemonic identity formations of gender and sexuality, race, and class. The interviews discussed here reveal some of the ways that dependent masculinity and American political culture limit the terms of political debate about an environmentally devastating practice. Appalachian problems are not typically understood as national issues. The politics of MTR make it clear, however, that the region is a crucial site in the ongoing construction of a particular definition of the American nation and American identity and in the making of a specific vision of the future. At the same time this identity is unsettled by the persistent traces of other ways of relating to nature and the land.

CHAPTER 4

ATVs in Action:
Transgression, Property Rights, and
Tourism on the Hatfield–McCoy Trail

THERE IS A STRIKING CONTRADICTION in the cultural politics of MTR: the same people who support and affirm the necessity of MTR also claim to love the natural beauty of the region. Although coal industry supporters may not see this as an inconsistency (because they apparently believe that reclaimed mountaintop mines are also beautiful), this contradiction nonetheless requires management in the public discourse. How can MTR possibly make sense in a place that calls itself "the Mountain State"? One of the entry points to understanding this inconsistency turns out to be the all-terrain vehicle (ATV), or four-wheeler. ATVs, like dirt bikes, are marketed nationally as a form of recreation. They are primarily used in rural areas, so their use frequently conflicts with environmental protection of fragile ecosystems. In the coalfields of southern West Virginia, however, ATVs are not used only for recreation. They are a part of everyday life and are used as transportation into the less accessible areas of forest and mountains. ATVs are integral to local land use practices that include habitual "trespassing" on corporate property while hunting, fishing, and gathering forest products. Ironically, ATVs can even be used in environmental activism.

More and more often these coalfield land use practices are coming into conflict with the authority of private property as MTR and other mining operations come into closer proximity to residential communities. At the same time, the local economy is being transformed, painfully, into a service-based economy, as mechanization increases and mining jobs dwindle. One of the most popular visions of West Virginia's future postcoal is an economy based on tourism. The eastern West Virginian highlands already support an active ecotourism industry, and in the coalfields tourism is seen as an industry that can potentially replace coal.

One way the state and local governments are pursuing this vision is through the Hatfield–McCoy Trails, a more than five-hundred-mile-long system of trails in the southern West Virginia coalfields. The trail system is nominally multiuse, but the presence of ATVs and dirt bikes tends to discourage hiking, horseback riding, or bicycling, and as the trail system's Web site makes clear, ATV users are the main target of trail marketing. The site features pictures of riders in full gear on ATVs and dirt bikes (Trails Heaven 2005). The trail system is a public-private partnership that takes advantage of the peculiar nature of land ownership in the coalfields, where two-thirds of the land is owned by corporations. It is meant to generate economic development in the form of lodging, restaurants, and other services for tourists who come to ride the trails.

As the term *public-private* suggests, the Hatfield–McCoy Trail System is fortuitous for the mining industry in several ways. The creation of an official ATV trail system to be used by recreational users brings a new level of bureaucratic oversight and management to the large corporate landholdings in southern West Virginia. Coalfield residents have used these corporate lands as a virtual commons for most of the twentieth century. The inception of the trail system, like the increasing prevalence of MTR in southern West Virginia, disrupts local land use patterns. Forest-based economic activity supplements diets through hunting and fishing, and incomes through the collection and sale of valuable plants like ginseng. By reserving the forest for recreational use, the new policing of corporate property encourages people to conform to a more modern, middle-class American lifestyle that does not include these forms of economic activity, while generating and enforcing new land use patterns that are not incompatible with MTR. Thus, serendipitously, the economic development desired by local governments is configured in such a way to help create a cultural context for MTR.

Transgressive Practices

I found out just how popular ATVs are in southern West Virginia during a school tour of a mountaintop mine. The company provided hot dogs, chips, and soda for lunch, served outside the prefabricated building where the mine office was located. We took our food inside, where we found a few offices on one side of the building and the rest of the space taken up by four or five large picnic-style tables. One side of the room held a row of soda machines, and the walls were covered with posters about safety procedures and other workplace rules. While we ate, we listened to a presentation on ATV safety given by a representative of the Mine Safety and Health Administration (MSHA). The MSHA officer asked the assembled students and teachers if they'd ever ridden an ATV. Everyone in the room raised his or her hand except me. The MSHA officer looked at me sitting there with my hands at my sides and said, "Well, I guess you're left out, aren't you?"

This experience indicates the ordinariness of ATV use in the coalfields. Previously, I was more familiar with another view of ATVs, exemplified by Craig, a white retired businessman who lives outside the southern coalfields. A nature lover, Craig enjoys hiking in the woods on his one hundred acres of property. Nearly all of it except for a few acres around the house is undeveloped second-growth forest. An environmentalist and outdoorsman, Craig values wild nature; he belongs to several organizations like the Sierra Club and the Natural Resource Defense Council. Craig's perspective on ATVs was the one I was predisposed to share. He hates ATVs. They are bad for the environment, they tear up vegetation and destroy creek beds. They cause noise pollution. Their riders ignore his no trespassing signs. When he sets up logs and other barriers on the paths, they simply go around, destroying even more vegetation in the process. For him, the ATV riders are ignorant of the harm they do to the environment, disrespectful of property rights, and aggressive in their willingness to trespass. By suggesting that there is more to the story than this, I am not discounting the very important ways that ATV use is harmful. Off-the-trail use does cause erosion. ATVs burn fossil fuels and emit harmful greenhouse gases. ATVs cause disruptions in neighborhoods and can lead to serious accidents and injuries, often involving young riders.

However, for many people in the coalfields, ATVs are a taken-for-granted part of everyday life. ATVs are ridden for both enjoyment and practicality and sometimes even serve political purposes. The discrepancies between these perspectives and Craig's point of view shed light on several local sites of contestation, including the meaning of environmentalism, private property, and freedom. Popular ATV use in the coalfields fits in with local understandings of rights in land and ideals of personal freedom and toughness. But local government and large corporate landowners have begun to transform the terrain of ATV use in the name of economic development. These changes also entail the rationalization and restriction of local land use patterns, the commodification of local historical specificity, and a homogenization of place that domesticates local transgressive practices, creating a postmodern space no longer incompatible with MTR.

Marjorie's case exemplifies the transgressive possibilities of popular ATV use. A white woman in her fifties, she cleans at night for a large local retailer. Marjorie is part of a group of extended family members who collectively own fifty acres on a mountain threatened by MTR. About twenty years ago, Marjorie's cousin Roger, the activist introduced in chapter 2, returned to the family homeplace and encouraged his relatives to join him in preserving this place as a part of their family life. Many of his relatives, including Marjorie, built cabins where they spend weekends and vacations. In the meantime, the mountains surrounding the homeplace have all been mined. Roger told me that their homeplace used to be one of the lower points in the area. Today it is the highest. The family's collective ownership of these fifty acres is all that is preventing their mountain from being flattened as well.

Mountaintop removal mine, Raleigh County, West Virginia, August 2004. Photograph by the author.

Roger regularly brings visitors to his homeplace because it is one of the few places on land owned by a private citizen where people can witness the effects of a mountaintop mine. Though as noted MTR is usually hidden from view, always just beyond the hillside that you can see from the road, Roger's place provides a rare panoramic view of "reclaimed" mine sites. This sad spectacle is an invaluable tool for the anti-MTR movement. The place is completely surrounded by MTR sites; sitting on the porch of Marjorie's small cabin, we could hear the sounds of an active mountaintop mine echoing through the trees.

Although Marjorie doesn't work full-time in the anti-MTR movement, she swears she will never sell out to the coal company. Owning and using her cabin on the mountain is itself a type of environmental activism. Her environmentalist ethic in this case resembles Craig's—they share the value of conservation. But interestingly, for Marjorie, ATV use is also part of her environmentalism—both in terms of the enjoyment of nature and in terms of political action. This contradicts Craig's ideas of the proper way to enjoy wild nature, where quietness in nature is a marker of difference from everyday modern life. Marjorie uses her ATV both for fun and transportation. The ATV helps separate leisure time from the everyday routine; when she visits her cabin, she rides up on her ATV, not in a car. It is also a way to experience nature while spending time at the cabin. She uses her ATV for exploring the woods and hunting, for instance.[1] She also mentioned another, politically significant, use of her ATV. Gradually, the mining around Marjorie's property has stopped and the property has been "reclaimed." Across the valley from the cabins, the mining was ongoing when I visited in 2004 and the noise of the machinery and the blasting echoed against the hillside. These mine sites are on private property and are officially off-limits to the public. But her ATV allows Marjorie access to this space.

The existence of the cabins is undoubtedly a sore spot for the coal companies that mine around the mountain. The place is thick with tension between coal company employees and environmentalists, with frequent confrontations on the road up the mountain. This site provides an unparalleled opportunity to witness MTR without company permission. The tours Roger offers to journalists, students, and interested people from around the world are a vital part of the anti-MTR movement, but the regular presence of people on the mountain causes trouble for the companies in other, more mundane ways as well. Marjorie described how the companies tried everything they could to keep her off their land, but she continued to ride

*Reclaimed flattop, Raleigh County, West Virginia, August 2004. Photograph
by the author.*

her ATV onto the MTR sites because she wanted to see what was going
on. She just let her dog run ahead and then told the guards, if they ques-
tioned her, that she was looking for her dog. She viewed the company's
claim to the right of private property as an attempt to keep what they were
doing secret, an illegitimate effort to keep her off the land. Marjorie's transgressive practices cross the barriers of conventional middle-class ideas
about private property, the proper enjoyment of nature, and the practice
of environmentalism.

Her use of ATVs is integral to her expression of connection with her
family's homeplace. She considers it safe to travel the area at will because
she knows the land by heart. And by keeping an eye on the MTR site, she
can keep up with the changes that the mining brings to the land as well. This
intimate knowledge of place is important in local definitions of freedom,
which depend on knowledge of the mountains, forest products, and an off-
the-grid ethic of self-reliance. Marjorie's use of ATVs directly contradicts

the stereotypical image of the antienvironmentalist ATV user. Craig sees the ATV users who trespass on his property as aggressive and ignorant, but Marjorie is purposely using her ATV to challenge and surveil the coal company. For her, the ATV is an element of her political self representing freedom and enabling her to make a claim on the land.

Private Property

The disjunctures between these different views of private property reveal the messy arbitrariness of cultural articulations between individualism, private property, and independence. Marjorie's ideas of private property are not the same as Craig's, which are not the same as those of the coal company. This reflects the problematic and contradictory nature of private property in the coalfields. Craig's view of his one hundred acres is a typical upper-middle-class view. He owns the property not for economic gain but for enjoyment, as a personal luxury. He values the woods and the freedom to explore them, as well as the relative quietness and privacy that his large amount of property provides. For him, there are a few correct ways to enjoy this space—mostly antitechnological ways, hiking, horseback riding, taking time to get to know the variations in vegetation, noticing the changes in season and so on. In many respects, this enjoyment of nature reflects the dominant environmental ethos of the Euro-American tradition; the forest represents an escape from the hassles of modern life. For Craig, the property functions something like an extension of his individuality. From Craig's perspective, when the ATV riders cross his property line, they not only damage the environment, they also offend his sense of privacy and use his land in illegitimate ways.

In the context of the corporate-owned land of the coalfields, ATV use looks much different. More than two-thirds of the land in the southern coalfields is owned by absentee corporate landowners whose interest in the land is purely economic (Appalachian Land Ownership Task Force 1983). West Virginia's low property taxes and coal severance tax system allow absentee corporations to hold the property in reserve; low property taxes make owning property inexpensive, and severance taxes, which are based on coal production, do not penalize companies for holding the property indefinitely (Haas 2008). What this means in practice is that the majority of undeveloped land in the coalfields has been owned by corporations for the entire twentieth century. And whereas Craig's notion of private property means that he sees his property as an extension of himself, the corporate notion of the sacred right of private property means something else.

Driving through southern West Virginia, I noticed a certain configuration of houses, roads, rivers, and railroads that repeats itself in so many places that I often had a sense of déjà vu. Small houses with narrow yards are squeezed between the road and the railroad, or between the road and the hillside, with a river or creek across the road. Chain-link fences separate faded plastic toys from passing cars and coal trucks. These small pockets of residential property tucked into narrow hollows are the remnants of coal camps. Sometimes, when the coal company finished mining in the area, they sold the small lots and the houses to the residents. Other times, they continued to lease the houses to the residents, reportedly at a very low rent. This almost feudal arrangement leaves the residents in either situation in an unenviable position. In the first case, if they own their small amount of property and their house, they stand to lose if the company decides to start mining in the mountains above their home or if underground subsidence or blasting damages their well or foundation.

In cases where a new mine or valley fill is likely to impact residents, the company often tries to buy back the property, which is what happened in Blair when the mountaintop mine was making life impossible there. But unless homeowners in this situation have improved their properties considerably, or can negotiate a good price, they are at risk of not being able to afford the same quality of house in another location. In the other situation, when people rent from the corporate landowner, they will simply be evicted. This had recently happened outside of the city of Logan, where an entire community, Rum Creek, was dissolved by its corporate landowner, the Dingess-Rum Land Company. The residents were offered pieces of land to rent in a nearby trailer park, but they had to be able to come up with the money to buy a trailer. This kind of eviction can also result from a mine-related disaster, such as the collapsed valley fill that washed away the community of Lyburn, in Logan County (OHVEC 2002). In this context the meaning of private property is quite different from the dominant middle-class view. Middle-class homeowners like Craig have an expectation that zoning laws and other measures will protect them from nuisances that lower the value of their property and impact their quality of life. But coalfield residents, who have often lived in one place for generations, have in practice fewer rights in their property and homes than the large corporations who own the majority of the land.

Roger recounted an anecdote illustrating this situation. An elderly couple who live in the area below one of the mines near Roger's place reported that the coal company would give them a phone call whenever they were about to blast. But instead of issuing a polite warning of the potential inconvenience, the company representative would instead tell them to leave their house during the blasting. Mining regulations do not allow companies to blast in a way that damages people's homes. But in the coalfields the property rights of the residents usually cede place to the right of the company to mine the coal. This story is indicative of the realities of living in the coalfields and the fragile security of homeownership when the economic imperatives of the coal industry determine local geography.

Nonetheless, coalfield residents have always had their own ideas about how the land around their homes should be used. With the majority of undeveloped land in the coalfields tied up in absentee corporate ownership, there are very few state parks or national forests. There are none at all in Boone County, which covers 503 square miles and has a population of 25,535, for instance, and only one, Chief Logan State Park, in Logan County, which is 456 square miles with a population of 37,718 (U.S. Census 2000). The small communities nestled in the hollows of the mountains are surrounded by "free" land. Most of this land is unavailable for purchase by private citizens or for protection by government. It cannot officially be considered wilderness or forest. In practice, however, there is a rich local culture of use. People have continuously used the forest in various ways despite its corporate ownership. They hunt deer, bear, raccoons, squirrels, and turkeys. They collect plants like ginseng and goldenseal. They fish. Children in the mountains have traditionally viewed the forest as their playground, where they build forts, hunt, fish, and explore freely.[2] Many people I interviewed talked about childhoods, either their own or their children's, in terms of running around in the woods or playing in the creek. For some, this is part of West Virginia's heritage, the forest and mountains and the freedom to explore them. Gina, a white anti-MTR activist, described this style of childhood as one of the things West Virginians stand to lose because of MTR.

Indeed, childhood freedom and the traditional relationship of mountain communities to the land is a central part of the argument that activists organize against MTR. Gina explained why she decided to move back to West Virginia to live on her family's land: "Because we're raised with different moral values here. [You] don't care so much about your PlayStation.

You care more about that field out there, or that trout stream back there, or that hillside up there, and you play outside and you're free." Linda, another white activist, cited the state motto:

> Mountaineers are always free, and that's because they can live on the mountains. And the only way they can stop them from being free is by destroying the mountains. And that's what they are doing . . . taking away our freedom. Because when they destroy the mountains, nobody can live here, nobody's going to be free. . . . No matter how much they've oppressed us, as long as we could live on the hills, even if this economy failed today, people could live. People here could make a living. But not if they destroy the mountains.

Linda's statement illustrates a sharp divergence between a predominant conception of freedom as the ability to participate in the market and a place-based version that represents freedom from the discipline of that national or global market. This argument signals the historical connections between restrictions on people's access to land and the expansion of market-based social forms (Thompson 1975).

At the same time as they stress the cultural specificity of West Virginia in their arguments against MTR, these activists have to deal with the negative side of Appalachian stereotypes. The ambivalence was also evident in a conversation with Laura, an educator and daughter of a coal miner, and her husband, Steve, who was not raised in the coalfields. We were discussing the difficulty of dealing with household garbage in rural coalfield counties where there is no garbage collection service. I told them that while living in Boone County, I found myself in the uncomfortable position of having to haul my garbage to the dump in the back of my station wagon. Laura laughed at this and said, "In the cities they have SUVs. We all have pickups." Steve noted that one way people deal with their trash is to make their own unofficial dumps. The problem of what to do with garbage raised the question of land use in general. Steve continued:

> In this particular area, people lived on these postage stamp [lots] and that stuff [garbage] piles up. People had no alternative . . . [and this] carries over into land use and so many other different facets of everyone's mentality. For the people who have been raised here over generations . . . a lot of this touches on . . . property. You

know, that whole mountainside back there . . . is all privately owned by corporations. Every single one of them has a no trespassing policy. No trespassing. No, you can't go up there, no, no, no, no. But there are no posted signs. I guarantee you all the people who live around here . . . the gentleman who lives next door has lived here for forty or fifty years; if he wants to go up there and hunt, he just goes, and there ain't anybody going to stop him because he's been doing it since he was a boy. And it doesn't matter that it's never been his property. There's that sense of that entitlement.

Laura, who was raised in a coal-mining community, interjected, "Well, because they don't allow you to own any of it, they force you." But Steve continued:

People have adopted certain behaviors and certain mentalities, about how they behave down here, and how they do certain things like land use that has been passed down from . . . generation to generation, even though they don't physically own the property. "What do you mean I can't go over here and deer hunt? What do you mean I can't go over here and cut firewood? We've been doing this . . . well, no, it ain't my property, but what do you mean, I can't?"

I responded that it sounded like people's habits of use gave them a sense of ownership. He continued, "I think it's deeper than just habits of use. These people in these coal camps . . . a lot of them have been there time out of mind. It's become a part of their ingrained functionality." And Laura replied, "Well, all they've ever known is these mountains and these rivers. What else other recreation is there?" She talked about the history of land ownership in the area. "My ancestors are traced back to 1610. Coming in here. They owned all of Logan, and the place over where Shoney's [restaurant] is, in that area called Black Bottom . . . and we've got the genealogy that says our family owned all of that, back to 1610. And you just wonder what happened." Steve explained, "And then the coal companies came in, in like 1880 or '90." But he also belittled this local sense of history, interpreting the repetition of stories like this not as evidence of their truth but as a sign of their inauthenticity. "It seems like almost every family down here, somebody says, 'We used to own all that, and they came in and my granddaddy sold it to them for nothing, you know, and now it's worth millions.' You know, it seems to be a common thread here."

While Steve and Laura share a sense of the history of landownership in the area, they seemed to have divergent views of what that history means for the present. Steve's representation of the local mentality, which reflects the perspective of a self-described outsider, seems to see the local habits of use as an indication of ignorance or a sense of inordinate entitlement. Laura defends local practices in terms of the structure of absentee corporate land-ownership and a (nativistic) morality of long-term habitation of the area.

When considered from the perspective of the hegemonic authority of private property, these local land use practices are problematic. ATV use fits into this pattern of highly developed habits of land use supported by very few legal rights to that land. For instance, Marjorie's attitude toward the coal company's property could be interpreted as a lack of respect for private property, as Craig interprets the trespassing of ATV users on his land. But as we've seen, her rejection of the company's rights is an overtly political act. When people ride their ATVs onto mining sites in defiance of a company's armed guards, they challenge the company's legal right as property owner to do anything it wants to the land. They insist on their right to know what is happening to the land, to witness, as stakeholders in the local environment. The people discussed by Laura and Steve who hunt and gather firewood in the coalfield forests are also arguably involved in a political act, a resistance to expropriation, although they may not articulate it that way. But Laura points in this direction when she justifies trespassing in her statement that "because they don't allow you to own any of it, they force you."

The ambivalence of local land use patterns was also evident when I visited a West Virginia history class at a coalfield middle school while on a tour of the area with one of my participants. I told them I was studying the coal industry and MTR. From the back of the room, one boy said he thought MTR was destroying West Virginian heritage. I asked what he meant by "heritage," and he responded, "The way we were raised, the way we grew up." When I asked him what he meant, he nervously explained, "We've got all kinds of stuff, there's so much stuff to do, every day you learn something new." Some of the other students did not agree. Several turned around to look at him and asked incredulously, "What are you talking about?" One girl said, "There is nothing to do here but ride bikes and ATVs." Another girl advised me, "Go tell the coal company, or all the rich people around here, the people that run [this place]. Tell them to come and talk to kids . . . about what kids want . . . we want something to do, we want a bowling alley like up in Charleston. We want a movie theater like up in Charleston. We don't want to have to go all the way up to Charleston to do anything."

Where some see the unique characteristics of the coalfields as an element of local "heritage," others seem to refuse this difference and desire a more mainstream, modern American life. The economic and cultural marginalization of the area leads some children to devalue the one widely available local resource available to them: forests to run around in and explore. The freedom that this space represents is devalued in two interrelated ways. This type of unstructured, unsupervised play is not in line with current hegemonic cultural norms. It is symbolically linked to the past, and it does not fit with American middle-class patterns of consumption.

Judging from my own experience growing up in West Virginia in the 1980s, the negative attitude of some coalfield children toward everything West Virginian is not new. The cultural marginalization of the local makes the natural wealth of the area seem uninteresting and negligible to some kids who want to belong to the powerful national culture they experience in the media. However, others grow up loving the freedom they experience and seem to understand on some level that they are experiencing something exceptional in modern America.[3] This Appalachian difference is grounded in a commitment to long-term inhabitation of the area. The phrase "I was born and raised here," sometimes followed by "and I want to die here," is used frequently by people to claim this nativistic status. This claim is an assertion of difference from the hegemonic American attachment to mobility. As Laura explained, "We're here [in this house] because we couldn't settle on buying property to build a home. Because [of] the [type of] home we wanted to build, well, Steve says we'll never get our money back. But my mentality is, 'I'm not going anywhere. Why do I worry about getting my money back?' I'm going to probably live and die here. It's very unlikely that we're ever going to leave."

This claim to a connection to place based on long-term occupancy is similar in structure to the claims of Native people to land. It draws on a tradition of place-based identity in opposition to a modern, abstract relation to land. The use of this nativistic metaphor in explaining Appalachian culture is problematic because while reflecting a marginalized experience common to many West Virginians, it also expresses a normative "white" American identity by eternalizing the existence of Euro-American settlements and erasing the original Native inhabitants of the Appalachian Mountains.

I examine these phenomena more closely in chapter 6 and the conclusion. Here, I am concerned mainly with the way that coalfield residents understand their usufruct rights to the land around them. Despite the history of absentee land ownership by corporations, coalfield residents have continually used the forest and mountain resources around them for hunting, gathering, and gardening in order to supplement their diets and incomes (Pudup 1990; Corbin 1981; Greene 1990). This taken-for-granted sense of a right to use the land around one's house is reflected in the comments of Shawn, a white activist. He made no mention of who owned the land as he detailed for me the ironies of MTR. State law strictly regulates the collection of endangered plant species like ginseng and goldenseal, forbidding the collection of whole plants and roots, for instance. These laws are meant to protect the species and their habitats. But MTR, which destroys all life on a mountain—including the ginseng and goldenseal—is not illegal. Shawn complained that these plants were becoming so rare in Logan County that "you have to go to Mingo County" to find them.

However, this coalfield culture of land use comes under attack both as a symptom of cultural backwardness and as a transgression against corporate property rights. Because the historical specificity of West Virginia is understood by the dominant culture as aberrant, this pattern of use is also understood as deviant, as Steve's comments illustrate. When this local land use is confronted by the hegemonic legal authority of private property, it is framed as a backward remnant of the hillbilly past, or as a reflection of a "coal miner mentality" of unreasonable entitlement. As mountaintop mining expands, and as coal seams near residential areas are being mined more often, this legal authority is becoming more obvious.

Shawn told me he thought the companies' intention was to depopulate the area.[4] He noted that due to subsidence from underground mines and other mining-related hazards, it was no longer safe for him to go into the mountains behind his home. This increasing lack of safety in the mountains was echoed by the MSHA officer who lectured the schoolchildren on ATV safety during their field trip to the mine; along with the advice about avoiding abandoned mine sites, he issued a general warning to stay off corporate property. Gina also told me how things had changed; when her son was seven, he was free to play in the woods around their house. Now that he's thirteen, it is no longer safe for him to do so.[5]

The sense of freedom represented by the ideal of a childhood spent running around in the mountains is also a site of cultural ambivalence.

Appalachian children have been used as symbols of the region's poverty and backwardness at least since the 1960s, when the political will for the War on Poverty was sparked nationally by the Charles Kuralt–hosted TV special *Christmas in Appalachia*.[6] Televised images of shoeless white children with dirty faces embodied the exceptional white poverty of Appalachia. This exceptional white poverty is one reason that Appalachia, which is a symbolically "pure" white region in the national imaginary, continues to be represented as America's Other (Klotter 1980). This image of Appalachian poverty as embodied by children continues today, with a "Santa Train" delivering Christmas gifts to the always needy children of Appalachia each year. The trope of needy white children in Appalachia was repeated in February 2009, when ABC aired a special titled *A Hidden America: Children of the Mountains*.[7] A barefoot child, playing in a mountain creek, hunting crawdads or collecting smooth rocks, can potentially index either nostalgic dreams of premodern freedom or nightmares of abjection and poverty. Of course, some of this distinction depends on whether the creek is clean or filled with acid mine drainage. In the context of cultural assumptions about Appalachian poverty and backwardness, these local land use practices signal abjection, not freedom.

At the same time as these practices exist in an uneasy cultural context, the literal terrain on which they occur is also uneasy. MTR is transforming the land that people are so familiar with, erasing whole places and ecosystems. What can ontological security mean when people begin prefacing their remarks with "Before the mountain came down"? As Shawn pointed out, MTR is entirely incompatible with local patterns of land use—it totally destroys the habitats of forest products and nullifies the specialized knowledge that nearby residents have of the land. Sometimes new ecosystems are introduced in reclamation—for instance, herds of cattle have been brought in to graze on the newly created grassland in places that previously only supported a few family-owned animals. MTR is fundamentally, basically, opposed to the coalfield cultural tradition of attachment to place. Several people suggested that the only ones who would soon remain in these transformed mountain places would be the elderly; everyone else would have to leave in the interest of their future, no matter how attached they are to their homeplaces, which might in any case be unrecognizably altered. This problem is exemplified by Roger's family's cabins. These cabins look like a great place to "get away from it all," relax in nature, and enjoy the old homeplace, where many of them were raised. I couldn't help but wonder, however, how much they could enjoy the place with the noise of the neighboring mine hammering through the trees and the constant reminders of what has been lost.

The Hatfield–McCoy Feud, Revisited

Into this context comes another transformation, the public-private partnership of the Hatfield-McCoy Trail System. For the first time, people attempting to hunt or fish on the land around their houses might be faced with a ranger charged with preventing trespassing on corporate property, keeping riders on the official trail, and making sure everyone has paid for a trail permit. As Ned, a white county official, told me, these rangers will at first issue citations but ultimately are authorized to physically remove trespassers from the property. Although riders pay to use the trails, the trail system is not expected to become self-supporting for some time. Right now the state subsidizes the trails in the name of the economic opportunities they promise. Permit prices are reduced for in-state users because their taxes are paying for the upkeep of the trails. In a sense, the right of coalfield residents to use the corporate-owned land is now quantified into a reduction of this ticket price. Even with reduced prices, local ATV users are averse to paying twenty-six dollars a year plus tax for the right to ride where they are used to riding for free. In addition, the purpose of the trail system is limited to recreation, and riding is confined to a specific trail. People riding ATVs while hunting, for example, are not going to accept being told where they can ride and where they cannot.

With the introduction of rangers and insurance policies, the Hatfield–McCoy Trails are bringing about the reprivatization of corporate property that has been used as a type of commons for more than one hundred years. According to Ned, who was involved in the creation of the trail system, it was difficult to convince the corporate owners to agree to the idea. I suggested that the idea seemed to be beneficial for the landowners, and he replied, "It took a long time to persuade [them] that this was good. It must have been two years we had to wine and dine people to get them to sign on. . . . When you bring in something new, it's hard to sway people's opinions on things. We're fortunate that our major landowner where the trails are is a lawyer. So he saw the writing on the wall, and the fine print that this would be very beneficial for his property." In this case, the major landowner is a wealthy West Virginian family that leases their property to a coal company for mining. Most of the land in question is owned by absentee land corporations with little interest in the small-town economies that the trail system is meant to grow.

The overt purpose of the Hatfield–McCoy Trail System is not about reasserting private ownership. But the creation of a bureaucracy to manage and protect the trails does in fact strengthen corporate land rights. As Ned described it: "It was a project that was started by [two men,] one was a consultant and one was a businessman, and they had this idea. . . . You know we had all this land in southern West Virginia just sitting here . . . [owned by] land companies, coal companies."

Of course, the land was not just sitting there but was being used in various ways by the people who live on or adjacent to it. But the activity that was taking place was illegitimate and not significant in market-based economic terms. This activity is not taken into account in the version of West Virginia's future envisioned by the founders of the Hatfield–McCoy Trail System. Their vision of the future includes rationalized, more "productive," land use. In this view, West Virginia postcoal will have a diversified economy based on tourism and economic development on reclaimed mines, including vineyards, fish hatcheries, lumber mills, prisons, golf courses, regional airports, and large consolidated schools, all made possible by MTR. Interestingly, in terms of fossil fuel dependency, one proposed development project in Mingo County is a NASCAR racetrack on a reclaimed mine site.

In testimony before a 2002 Senate subcommittee hearing about proposed changes to the Clean Water Act, Mike Whitt of the Mingo County Redevelopment Authority and "Chief Trail Scout" of the Hatfield–McCoy Trails argued:

> With mining, Mingo County is diversifying the economy. . . .
> Because of mining and development sites created by mining,
> we have been able to create good jobs in the industries of wood,
> aquaculture, agriculture and recreation. The Mingo County Board
> of Education has established a Horticultural Curriculum through
> the use of our agriculture demonstration project. By growing
> excellent Arctic Char from mine water, we have created a new
> industry in southern West Virginia. We anticipate the county
> school system will add an Aquaculture Curriculum as a result of
> our fish hatchery, grow out facilities and proposed fish processing

facility. Without mining, these new jobs and economic opportunities would never have been possible in southern West Virginia! (U.S. Senate, 22–23)

The Hatfield-McCoy Trail Scouts Web site introduces Mike Whitt as follows:

> Mike Whitt is the only chief Trail Scout who lives in the Hatfield–McCoy project area. He is the Executive Director of the Mingo County Redevelopment Authority in Williamson. A former state legislator, Mike enjoys an excellent relationship with state and local elected officials. He has served as the Hatfield–McCoy liaison with the landowners and has successfully encouraged them to sign licensing agreements. Lately, he's also been serving as the Authority's interim director, filling in at a critical time and working long hours to get the trails open. Like all southern West Virginians, Mike seems to have been born on an ATV. He rides frequently to the top of a reclaimed surface mine where he operates a demonstration agriculture project growing grapes and other fruit in old coal gravel. The chief Trail Scouts and their wives toasted the opening of the trails with Mike's strip mine wine. (Trail Scouts 2005)

On the same page Mike Whitt personally comments, "Welcome to southern West Virginia where the mountains are beautiful, the people are warm and friendly, and the trails are nothing short of Trails Heaven. We hope you will enjoy our hospitality and the land that we love so much, and that you'll come back again and again to have fun here."

The Hatfield–McCoy Trail System's stated purpose is to create economic growth. But it also disciplines land use in southern West Virginia while ensuring safety in increasingly dangerous environmental conditions. It brings a new form of biopower to bear in the region by regulating residents' relationship to the land. The trail system contributes to the rationalization of local space by attempting to limit ATV riding to recreational use on groomed official trails (Thongchai 1994). It works seamlessly with other envisioned development in the region such as Mike Whitt's demonstration vineyard. As the Trail Scouts site makes clear, the Hatfield–McCoy Trail System is aimed at a national audience: "There are now more than 12,000 Hatfield–McCoy Trail Scouts all across the nation" (Trail Scouts

2005). The people who come to West Virginia to ride the trails and enjoy the local hospitality are not likely to be critical of the changes MTR has brought to the local environment. Certainly, the popularity of an ATV trail is not dependent on the same degree of natural beauty, quietness, or isolation that a hiking trail requires. While riding ATVs recreationally, people are primarily enjoying movement through space and concentrating on navigating the terrain immediately before them, especially when they are not intimately familiar with the land. In this sense, riding an ATV is like driving a car. Unlike the place-based patterns of land use that are common among coalfield residents, recreational ATV riding is not incompatible with MTR. As of 2005, about 80 percent of Hatfield–McCoy Trails users were from out of state.[8] During the lecture on ATV safety during the mine field trip, the children and teachers responded derisively to the idea of paying to use the trail. Besides the problem of the fee, the trail is not compatible with the main uses of ATVs in the coalfields. Many of these uses have elements of practicality: hunting, gathering forest products, transportation to specific places. Definitely, people also use ATVs for fun, but since this recreation happens in a particular, familiar place, people are able to do both: enjoy the ride and enjoy nature.

This use of ATVs in specific places is tied up in many ways with coalfield ideals of toughness and self-reliance. Gus, a retired miner and resident of Blair, told me a story similar to Marjorie's:

When they strip they won't let you back on it. At least back here, they won't let you back there . . . they've got a security guard and he'll try to take you off down the other side where the guard shack is. I lost a dog down through there the other day and I took my four-wheeler down on that strip, went out through yonder hunting for my dog, and [the guard] caught up with me and he said, "You ain't supposed to be back there on those four-wheelers." And I said, "I lost my dog; I'm hunting for my dog." He said, "I've got to take you off back to that guard shack back yonder." I said, "You ain't taking me nowhere." I said, "I'm going back the way I come." I had another dog with me and I put that dog on behind me on my four-wheeler and I come back over this way. Never did find my other dog.

He also recounted a story about his grandson, who was riding a motorcycle on the mine when the company guards started chasing him. "My grandson rode back there on a motorcycle and [the company guards] got after him, and he took off on that motorcycle, and he come to a dead end, and they thought they had him trapped. He just jumped off that motorcycle and took it down the mountain. They thought they had him trapped, but he just [went] down over that mountain [walking his bike]. Once he got over the hill, that was the end of them, they weren't going after him."

These stories of standing up to company guards and bravery in ATV and motorcycle use demonstrate local knowledge of the land. Gus and his grandson have an edge on the company guards they face down; they have an intimate knowledge of the land around them. They know more ways around the mine than the company guards do, and the grandson's escape reveals a body-level familiarity with the mountains that the guards may lack. Part of the enjoyment of riding ATVs and motorcycles is this very familiarity. In this way, ATV use in the coalfields crosses barriers, both in terms of property lines and of ideas about the proper way to enjoy nature.

As Gus's stories illustrate, the coalfield ideal of toughness is expressed in part through ATV use. By using the name Hatfield–McCoy, the trail system signals a similar ideal. The Hatfield and McCoy feud of the late nineteenth century was one of many such violent episodes that went on throughout the nation. However, this particular one was called a feud in a reference to the supposedly Celtic or Anglo-Saxon roots of the participants. It was made famous by the book *An American Vendetta*, by T. C. Crawford, published in 1888 (Batteau 1990; Blee and Billings 1999). The incident continued to titillate, and other titles illustrate the kind of angle authors usually took: Appalachian romance author Richard Fox wrote *Devil Anse, or the Hatfield–McCoy Outlaws, a Full and Complete History of the Deadly Feud Existing between the Hatfield and McCoy Clans,* in 1889. In 1946, the topic was revisited in an article by Dan Cunningham called "The Horrible Butcheries of West Virginia," published in the *West Virginia History Journal* and in the book *Their Ancient Grudge,* by Harry Harrison Kroll (West Virginia Archives & History 2006). The Hatfield–McCoy Trail Scouts site emphasizes its connection to this feud by the inclusion, at the bottom of the "Meet the Chiefs" page, of nineteenth-century photographs of "The Original Chief Trail Scouts" William "Devil Anse" Hatfield and Randolph "Old Ranel" McCoy.

Despite the association of the famously feuding families with the negative Appalachian stereotypes of isolation, violence, backwardness, and clannishness, the Hatfield–McCoy Trail System exploits the lurid image of tough, violent mountain men as a marketing angle. Like Julia Bonds' reference on C-Span to hillbillies, quoted in chapter 1, the feud contextualizes the area in the national imagination. It may create an air of exoticism or simply provide name recognition. The Web site employs ambiguity in its branding strategies, clearly denoting "primary" or "lead" in the phrase "Chief Trail Scouts" but at the same time evoking the American cultural imagination of Native Americans and an aura of the frontier. This underlines the articulation of white Appalachian identity with concepts associated with Native Americans. But as with the reference to hillbillies, trail boosters and businesses associated with the trail have to try to distance themselves from the negative or scary valences of these references.

Publicity for the Hatfield–McCoy Trails domesticates Appalachia's reputedly dangerous past and commodifies it. Outside the city of Logan, visitors can stop in at the Hatfield–McCoy Market next to Walmart, where along with their gas they can purchase T-shirts decorated with feuding cartoon hillbillies. The trail bureaucratizes the local practice of transgressing property lines and hires rangers to police it. The trail repackages coalfield land use patterns and regional identity in commodified form. Like Celebration, Florida, and the Baltimore Harbor, the Hatfield–McCoy Trails and related tourist services create a safe place to experience a sanitized, shallow version of history (Gottdiener 2001; Harvey 1990a; Ross 2000). In marketing the amenities available to trail riders, trail boosters and associated businesses play up local specificity as friendly, old-fashioned southern hospitality. The marketing of the trail system removes the teeth from Appalachian stereotypes at the same time as it plays on the excitement of entering dangerous hillbilly territory. The feud is emptied of its negatively charged associations with violent mountain men, as well as its associations with isolation, poverty, and exploitation by the coal industry.[9] In advertising copy that evokes the name of the famous feud while simultaneously assuring visitors that they will be safe, appreciated, and well cared for, Appalachian regional specificity is reduced to an empty marketing signifier while the region is made over as part of a homogeneous American nation.

Coalfield ATV use, as we've seen, is a transgressive practice. It is involved in the creation of local cultural specificity through idiosyncratic land use patterns and through its facilitation of the ideal of personal freedom. It presents a challenge to corporate landownership and enables ordinary people to surveil coal-mining operations against the companies' wishes. But MTR is rapidly changing the terrain on which this activity takes place. Most coalfield land use practices, which build and rely on an intimate knowledge of specific places, are utterly incompatible with MTR. MTR is making local land use practices obsolete; whole areas of the mountains are no longer safe, entire habitats are destroyed, along with all the resources of the forest. Intimate local knowledge of the land becomes passé as "the mountains come down." By flattening the mountains, MTR reduces the geographical specificity of southern West Virginia. And local government is pursuing an agenda of economic development that works in harmony with MTR to create a modernized, more homogeneously American West Virginia.

CHAPTER 5

Coal Heritage/Coal History: Appalachia, America, and Mountaintop Removal

Fire in the hole! Fire in the hole!

—U.S. soldier in Iraq, blowing up an abandoned car suspected
of belonging to insurgents. ("The Baghdad Bomb Squad")

EVER SINCE THE "GREAT LAND GRAB" of the late nineteenth century, southern West Virginia has been defined as a coalfield. The coal industry's priorities have determined the local economy, culture, and geography. While some see the way MTR is rapidly transforming Appalachian landscapes and communities as economic progress, activists counter that this extensive resource extraction is incompatible with nearby communities and ways of life. The politics of MTR lay bare the uneven roots of America's postindustrial economy, raising questions about the meaning and direction of the American nation and Appalachia's place within it. In the national media, images of Appalachian poverty and environmental devastation periodically appear, disrupting national progress narratives. Appalachian differences unsettle notions of a uniform American modernity (K. Stewart 1996). These images often evoke scale-bending comparisons to the so-called Third World, in a spatial politics that reasserts difference and conflict into abstract national space. The question of how Appalachia fits into America also haunts coalfield communities: are the sacrifices endured by the region evidence of their exclusion or of their belonging?

As a marginalized region, Appalachia is excluded from the American mainstream, while at the same time it represents for many the "heartland," or the "real America." The billions of tons of coal taken from the mountains have cost the lives of countless workers and caused untold damage to the environment. In that sense, the Appalachian coalfields are a national sacrifice zone (Fox 1999; Kuletz 1998). Simultaneously, however, the place has served as a screen for the projection of some powerful elements of the national symbolic. In historical and literary accounts, it has been a location for an individualistic, quintessentially American struggle against nature. The region has been imagined as a cultural reservoir of "unadulterated" Anglo-Saxon cultural heritage, a source of true patriotism, and a site of degeneration (Batteau 1990; Foster 1988; McNeil 1995). These contradictions run right through the region, marking it as an uncanny place where the American paradigm of economic development is haunted by the specter of failed progress, precisely in the heart of American national space.

This chapter maps two accounts of coalfield history that reflect and attempt to manage these contradictions. The divergences in these accounts remind us that popular memory is a political practice; the "past-present relation" is the object of this practice (R. Johnson et al. 1982, 211). The first is the coal heritage movement, represented here by my reading of two West Virginia tourist sites—the Coal Heritage Museum, in Madison, and the Exhibition Coal Mine, in nearby Beckley—and of regional development plans related to coal heritage tourism. The second account of coalfield history is a historical preservationist movement focused on protecting the site of the 1921 Battle of Blair Mountain from MTR. Both the coal heritage movement and the historical preservationist movement are invested in the survival of the material traces of memory. For the heritage movement, everyday artifacts of coal camp life come to represent disappearing communities and lifeworlds. For what I am calling the Save Blair Mountain movement,[1] the potential destruction of the battlefield by MTR indexes a long history of more or less figurative bulldozing of mountain communities in the name of outside interests. Both of these visions of the coalfield past interweave "nostalgia and critical memory" (Baker 1999) in order to tell a particular story. As Doreen Massey notes, "the definition of any particular locality will . . . reflect the question at issue" (Massey 1994, 139). These two representations of the past are involved in competing scale-making projects concerned with the meaning, memory, and future of Appalachia as a place in itself and in terms of its belonging to an abstract national space (Lefebvre 1991; N. Smith 2004). These divergences are also reflected in the institutional frameworks that support each of these accounts of the past: the National Register of Historic Places and the National Heritage Areas Program.

A Contentious History

In the most general terms, the history of coal mining in southern West Virginia goes something like this: in the late nineteenth century, growing national demand for coal brought new interest in the mountains of southern West Virginia. Large amounts of coal-bearing land were purchased from the local white inhabitants, sometimes under dishonest conditions (Caudill 1962; Eller 1982; Gaventa 1980). Railroads and land corporations have been able to make untold profits off the leasing of these lands and mineral rights over the course of a century. Coal companies needed a larger, more concentrated workforce than the existing rural population could provide, so they set up coal camps. They recruited workers from Europe and the American South and provided them with minimally acceptable housing, small, four-room houses, usually built in a hurry out of cheap, green wood—not built to last.

Worker safety was not a major concern for coal companies in the days before the union. Men were "cheaper than mules," easily replaceable by more migrants from the South or European immigrants. In those "hand loading" days, mining was an extremely dangerous, labor-intensive industry. The billions of tons of coal that left the mountains to fuel the national industrial economy were extracted by the intense, painful, repetitive, dangerous physical labor of hundreds of thousands of men (Corbin 1981; Eller 1982).

In those days, as now, Appalachian regional identity was characterized by a contradiction. During the First World War, Woodrow Wilson exempted coal miners from the draft, declaring, "Scarcity of coal is the most serious danger that confronts us" (quoted in Shogan 2004, 50). Like soldiers at war, miners risked dying in cave-ins or explosions, losing limbs, or being poisoned by gas, but the coal the miners produced literally fueled the war effort. In 1918, the *Mine Workers Journal* described the deaths of thirteen miners in an explosion in these terms: "These local boys died in the interests of democracy, they were exerting their manpower in the production of coal with which to help win the war" (quoted in Shogan, 50–51). In this sense, mining coal was an expression of citizenship, an essential service to the nation. Not long afterward, however, Appalachian cultural marginalization asserted a spatial discipline on the disorder of industrial capitalism (N. Smith 1992). During the 1920s Mine Wars, national newspapers announced coal miners' essential difference, referring to unionizing miners as "strike feudists," for example (Batteau 1990, 111). Portraying miners as "Primitive Mountaineers," or "sixteenth century fauna" reaffirmed that this violence stemmed from the region's peculiarities, not from a national class struggle (112). Meanwhile, at the urging of the industry, West Virginia established a state police force to aid private mine guards in maintaining order in the coalfields (Williams 2002, 261). The workers had no real voice in state or federal politics. It was not until the 1930s that either the mainstream national press or the federal government recognized the miners' struggle as a "serious matter" (Batteau 1990, 119).

In what is known as the Coal Boom era, coalfield communities were bustling places. The large labor force required by the industry in the 1930s, 1940s, and 1950s led to a population boom, and the company towns thrived; department stores, restaurants, and movie theaters all sprang up to share in the profits of the coal boom. Year after year, the industry employed up to a hundred thousand workers. In the 1950s, mechanization and the beginning of surface mining reduced the need for labor, and people began to leave.

These changes affected miners' bodies, working lives, and communities in positive and negative ways. The work became physically easier, with machines like the continuous miner, mechanized loaders, and belts. Workplace regulations and safety equipment saved lives. Mechanization of underground mining increased the incidence of silicosis, or black lung, because the equipment pulverized the coal more effectively than the earlier hand methods (Derickson 1998, 139). Surface mining was initially unregulated, and these "shoot and shove" mines left scars all over the landscape, sparking an environmental movement that forced the federal government to intervene (Montrie 2003). The coalfields of McDowell County were mined out, and Welch was decimated. McDowell County, once a symbol of wealth, is now legendary for its poverty. In 2000, almost 40 percent of children and over 30 percent of the entire population in McDowell County were living below the federal poverty line (U.S. Census 2000). Madison, the county seat of Boone County, lost its movie theater, department stores, and other businesses, leaving its main street storefronts deserted or occupied by flea markets and antique stores. In 2004, the only restaurants in Madison were national fast-food franchises, with the exception of a Chinese restaurant and a barbecue sandwich shop located in a gas station. At the same time, the industry has gained advantages in collective bargaining, forcing contractual agreements that have loosened the UMWA's control of the coal production climate. Remarkably, in a part of the world where children grow up knowing the lyrics to "Sixteen Tons" and the other meaning of the word *scab*, antiunion forces have made the phrase "union-free" part of a new lexicon of coal mining.

In January 2006, the results of these shifts became obvious, even to the normally uninterested national media. Fourteen nonunion miners died in accidents that month with the deaths at Sago followed by two more at Massey Energy's Aracoma coal mine. These tragic mining deaths make the sacrifices of mining communities apparent on a national level, but inside mining communities these sacrifices are always apparent. In any given year, it is not unusual for ten men to die in mining accidents in West Virginia. In 2008 there were nine fatal accidents; 2006 was a really bad year, with a total of twenty-five deaths (West Virginia Office of Miners' Health, Safety and Training 2009). The January 2006 deaths generated new interest in how mining safety is regulated and revived the public's outrage at the token fines with which companies are penalized for safety violations that can cost lives. Critics of the industry often remark that companies simply consider them part of the cost of doing business (Burns 2007). What coalfield residents know only too well—that miners' deaths are also part of the cost of business, not only for the coal industry but for the entire national economy—is less likely to be discussed outside the coalfields.[2]

Coalfield communities have shrunk with the increased mechanization of mining, but the effects of mining on those that remain seem to have increased. Newer intensive underground mining methods such as longwall mining increase the risks of subsidence, which can destroy water wells and structures on the ground above. Instead of the small "tipples," or coal preparation plants, that used to dot the landscape and whose emissions covered nearby houses with dust, the companies now operate fewer but larger preparation plants with concentrated effects, such as the one that dominates downtown Keystone, in McDowell County. MTR enables an unparalleled "efficiency" in mining coal, but as noted it literally transforms the landscape, destroying forests, aquifers, and the headwaters of streams. A working mountaintop mine makes nearby communities uninhabitable, creates new flood hazards, and generates other unpredictable environmental risks. MTR is worlds away from the underground coal mining of the Coal Boom era. Instead of company towns, it builds empty space and destroys communities. Instead of generating a rich "coal culture," it erodes the cultural specificity of the mountains.

The uncertainty of the coalfield future has elicited an increasing interest in the preservation of its past. The Coal Heritage Museum houses many of the artifacts representing these lost or disappearing cultures of coal. These ordinary objects become souvenirs of a lost world, as mechanization severs mine labor from its landscape (S. Stewart 1984). "Coal heritage" offers people a way to hold on to a rapidly disappearing coal culture, but it is shaped by cultural, economic, and political forces that encourage selective forgetting (Connerton 2006). Indeed, heritage cannot properly be termed history at all; rather it is "a publicly instituted structuring of consciousness" that "functions by excluding traditions it cannot incorporate" (Bommes and Wright 1982, 266). Reflecting the interests of some of its financial and institutional backers, the "coal heritage story" locates coal mining culture within a narrative of American technological and social progress. This liberal progress narrative is part of the habitus of American modernity, and it inhabits and shapes the coal heritage story without being spoken (Bourdieu 1998). Heritage shapes history into a conventional narrative marked by a compelling simplicity; the "coal heritage story" substitutes a national allegory of development for the messy regional history of coal (Baker 1999, 264; Berlant 2008).

Meanwhile, a few people up on Blair Mountain, located only a few miles from Madison, have spent years scouring the forest for bullets, guns, and other evidence of the 1921 Battle of Blair Mountain, when union miners marched on the antiunion stronghold of Logan County and federal troops were sent in to stop them. The activists have been working to get the battlefield, and the mountain, placed on the National Register of Historic Places. The site was listed on March 30, 2009, but was almost immediately delisted (Ward 2009a, 2009d). These efforts were fueled in part by the desire to save Blair Mountain from destruction—if coal companies prevail, much of the mountain will be removed in order to get at the coal beneath. All material evidence of this episode in American labor history, including the very land where it happened, is in danger of being erased by MTR. The Save Blair Mountain movement represents the eruption of critical memory against the grain; blasted and unsettled by MTR, Blair Mountain is literally the "unquiet earth" (Giardina 1994).

These two projects are oddly disconnected, despite their apparent affinities. Working under the aegis of local governments, the coal industry, and other businesses, coal heritage writes West Virginia history into a national story of progress, culminating in a cleaned-up, postindustrial economy, complete with the ultramodern mining technique of MTR. In this narrative, coalfield hardships represent a necessary sacrifice for the national interest. But according to the Save Blair Mountain movement, MTR is simply the latest form of hyperexploitation of a historically marginalized place. These preservationists' activism is simultaneously a fight for the environment and for self-determination, and in their view it fits perfectly within the history of American labor activism that the Battle of Blair Mountain symbolizes. Their project is part of a historical struggle for coalfield residents to claim the rights of full American citizenship, including the right to equality in difference. In their insistence on remembering the conflict between local and national interests, the Save Blair Mountain movement mobilizes a spatial politics of difference against the push of national homogeneity.

In the United States, national identity, racial identity, and class distinctions have historically been maintained through hegemonic standards in material culture (Heneghan 2003; Moskowitz 2004). Material culture is a primary means for the expression of both belonging and difference (Gaytan 2010). Prior to the ascendancy of coal, the distinctive material culture of Appalachia and its distance from the market economy were mobilized in the region's marginalization and were taken to indicate backwardness and poverty rather than an alternative way of life (Precourt 1983; Semple 1901). Since the rise of the industry, images of coal camp life have been associated with a degree of hardship, degradation, and exploitation that does not conform to ideals of American national identity. But national belonging can also be compelled or performed through discourses of sacrifice for the nation (Naono 2010). When the environmental and social damage caused by mining is read through a nostalgic lens of sacrifice for the nation, Appalachia's marginal status as a natural resource colony ironically provides coalfield communities a way to claim a core national identity. The hardships of mining can be read as evidence of their patriotic devotion to America and their central role in the national economy.

The coal heritage movement contains multiple projects reflecting different interests and desires. To a large degree, coal heritage is a popular history project. The artifacts collected in the Coal Heritage Museum are donated by miners and their family members who want the coal-mining life to be remembered and honored. The people who visit the museum and other heritage sites are frequently the children of miners, curious about their parents' lives and nostalgic for their childhoods. The volunteers and tour guides working in these heritage sites are likewise frequently personally invested in remembering all aspects of the coal-mining life. But in its capacity as the great hope of economic boosterism in the southern West Virginia coalfields, coal heritage resolutely focuses its attention on a fuzzy past, encouraging a nostalgic view of the hardships of early coal-mining culture. This is the vision of the local elites for whom coal heritage is an economic development initiative. This "coal heritage story" locates coal-mining culture within a narrative of American technological and social progress. Because coal heritage focuses most of its attention on the issues of the past, it distracts attention from the problems of current mining practices. The coal heritage story creates a context in which MTR represents the latest in a steady march of labor-saving mechanization in mining, which now offers the region the possibility of transcending its economic and cultural marginalization. The coal industry, in this story, has not only created the coalfields through its economic activity in the region, it has given the region the potential for overcoming this resource colony status in the form of coal heritage tourism.

Heritage and History

The history of the southern West Virginian coalfields includes some of the bloodiest events in American labor history. Armed thugs hired by the coal companies kept de facto martial law in the coal camps during the struggles for unionization of the early twentieth century. Sid Hatfield, the pro-union sheriff of Mingo County, was assassinated. Mingo County was known as "Bloody Mingo" because of the frequent gun battles between union miners, company guards, and "scabs." But coalfield residents also remember the first half of the twentieth century quite nostalgically as a time when company towns were the center of the world, with company stores, company entertainments, and company houses. People remember a rich culture of mining, represented by objects such as the small metal disks called "chits" (that miners used to mark the loads of coal they had mined when they were paid by the ton), carbide lights, and picks and shovels.

Reflecting local interest in this time, and the hope that others will be curious, the southern West Virginia coalfields themselves have been designated a National Heritage Area.[3] A series of roads through the area has been designated the "Coal Heritage Trail," a national scenic byway.[4] In Madison, local economic boosters developed the Coal Heritage Museum to accompany the annual Coal Festival. Yet the nostalgic face of coal heritage seems out of sync with the complex unfolding history of the coal industry in southern West Virginia. In the woods on Blair Mountain, Shawn searched the loamy forest floor for the rusty bullet casings that marked the terrain of the battlefield. As I scrambled after him, through the fallen leaves and around the trees, the kindred concepts of heritage and history suddenly seemed very much at odds.

Although the terms seem practically interchangeable in everyday popular use, "heritage" evacuates the politics from history. History as seen from within a heritage framework evokes nostalgia, and the violence and inequality of historical struggles and contemporary conflicts become very hard to represent in such a context (Bommes and Wright 1982; Levin 2007). The plantation tours of the American South offer an example of this heritage logic. The tours usually reproduce a racist, white-centered story of pre–Civil War plantation life, including slavery. Those few plantation sites that attempt to represent slavery and plantation reality in an antiracist way struggle to do so at the same time as they pursue their main mission, which is presenting a nostalgic view of white "plantation heritage" for mostly white tourists (Eichstedt and Small 2002). Coal heritage similarly cannot afford to focus attention too clearly on the realities of coal mining labor history. To do so might implicate the present union-busting activities of companies like Massey Energy in a way that would defeat the purpose of coal heritage—to attract tourists. This hoped-for postcoal tourist economy relies on a safe, pleasant packaging of coalfield history.

Thus, while the Save Blair Mountain movement struggles to preserve the history of a significant event in labor history, the coal heritage movement "writes the revolution as a well-passed aberration" (Baker 1999, 264). Coal heritage boosters commodify heroic stories of the sacrifices of early coal mining, along with images of a mountain/frontier culture, as part of an attempt to refigure an industrial wasteland into a postindustrial space of entertainment, nostalgia, and consumption (Bommes and Wright 1982; Zukin 1991). But the conflicting interests within the coal heritage movement highlight the disjunctures between the experience of people displaced by economic and environmental upheaval, people increasingly marginalized in the national economy, and the national progress narrative that shapes the coal heritage story.

Museums and memorials often reflect the tensions inherent in preserving the memory of historical oppression and resistance from within a social and cultural formation shaped by that oppression (Coombes 2003; Weissberg 1999). In Germany, struggles over representations of the Nazi holocaust highlight the difficulty of balancing the requirement to remember past horror with the need to define the present community in some bearable form. This struggle is even more pronounced in the case of democratic

South Africa, where the divisive legacy of apartheid remains embodied in national memorials and museums. These museums and memorials have become focal points in struggles around the representation of the nation, its past, and its citizens. In both cases, cultural memory is shown to be a conversation and a compromise between different agencies, audiences, and forms. Visitors with different perspectives read museum displays differently, while the designers of these repositories of memory shape their work with an anticipation of their audience. Similarly, I am interested in the stakes involved in these different memories and representations of the coalfield past, and, in light of current environmental and social transformations in the coalfields, their implications for the future. For a region constituted by marginalization and conflict, cultural memory is an object of struggle in competing efforts to define the place and the terms of action.

The highly publicized mining deaths of 2006 are grim evidence that the hardship and struggle of coal mining is not a thing of the past. Massey Energy Company was the largest coal producer in West Virginia in 2005 and has largely succeeded in its purported mission of destroying the UMWA in the state. For many environmental activists and organized labor supporters in West Virginia, Massey is public enemy number one (Ward 2001, 2005). The company's heavy reliance on subcontractors and subsidiaries helps it to hide its true safety and environmental record from the public. The February 1, 2006, death by fire of a bulldozer driver on a Massey mountaintop mine evoked local suspicions of the company. A local TV news program reported that Massey employees prevented first response emergency rescue teams from reaching the accident scene in a timely manner.[5] The report indicates the deep distrust of Massey that exists in local communities. Massey has been responsible for some of the worst environmental violations, including slurry spills and collapsed valley fills, in recent memory and has apparently turned back the clock on working conditions; as noted, the company reportedly routinely forces miners to work overtime, as long as it takes to finish the task at hand, or lose their jobs. Massey miners are not given meal breaks during their shifts but are told to eat while they work. Laura told me that things were getting so bad, back to pre-union conditions, that soon the miners would "get organized" again. The increasing prevalence of nonunion mines represents a significant shift in mining culture that disregards worker safety and rights in favor of profits and does not accord with the story of coal heritage (Ward 2006c, 2007, 2008d).

Two Sites in Coal Heritage

People in West Virginia are well aware that coal will not last forever. Some estimate that coal will run out in as soon as fifteen years; others are more relaxed, and think in terms of multiple generations.[6] But there is a general movement in West Virginia to begin thinking of the economy postcoal. Ironically, as the mountains come down, tourism is a key part of this post-coal economic vision. Winding roads through mountainous terrain in the heart of the coalfields are reconfigured as a Coal Heritage Trail. Local business leaders optimistically hope that rechristening the roads and offering maps and tourist information online will bring the curious to the area to spend their money.

The Bituminous Coal Heritage Foundation Museum in Madison was created as part of Boone County's annual Coal Festival. Clearly, the major purpose of the collection, for the donors, is to honor the memory of coal miners. It offers miners and their families a place to put the memorabilia of their working lives, a public place where their work can be recognized as important. Lunch buckets, different kinds of helmets, tools, union wage agreements, newspapers, chits, and scrip, when removed from someone's attic or closet and placed in a public setting, become more than personal keepsakes; they represent a lifestyle and a community. Many items are marked by cards noting who donated or loaned them. This embeds the museum in local social networks; these names are likely to be familiar to many visitors to the museum. At the same time, the "metonymic power" (Coombes 2003, 88) of these ordinary objects writes certain people and perspectives out of "coal heritage."[7]

The displays are loosely organized around several themes: the early "hand loading" era of underground mining, the union (which as noted itself increasingly seems to be a relic of the past), and developments in mining equipment, especially safety equipment. These grouping points allow the museum displays to be read as reflecting "progress." At the same time, however, the museum seems haunted by the sacrifices of miners. An underlying effect of mourning suggests the breakdown of the progress narrative. One vitrine is labeled "Memorial Section" and includes bullets from the Battle of Blair Mountain. White flags, each with the black silhouette of a helmeted miner, hang from the ceiling, commemorating the annual festivals. A large placard bearing the names of West Virginia miners recently killed at work is prominently displayed in front of the entrance.

The only other display to address current mining practices is a poster titled "Boone County's Coal Heritage," where sepia-toned pictures of past mining operations are placed next to a picture of a dragline, the giant earthmover used in MTR. In the lower left-hand corner, the poster reports: "Surface mines, like Hobet's large mountaintop removal mine, produce one-third of the county's coal. With its huge 180 ton bucket dragline (living room sized) removing the rocks above the coals, Hobet mines five seams." The short text emphasizes the impressive technology of MTR and minimizes its destructiveness by referring to the mountain that is removed as the rocks above the coals. This brief mention of MTR establishes underground mining as the unmarked subject of "Coal Heritage." Indeed, MTR is a constitutive force in coal heritage; its massive impersonal technology, which renders workers obsolete and communities uninhabitable, makes history of the artifacts documenting the lives of miners and mining communities. On one level, the dragline might represent a technological rescue from the hardships of the early mining days. But the small, quiet presence of the dragline in the museum also hints at a new kind of sacrifice demanded of Appalachia. This reflects the difficulty of representing something called coal heritage while coal-mining history is still being made. The heritage movement, which seeks to benefit economically from potential tourist interest in coal culture, is constantly faced with the coal industry's continuing extraction of coal from the region.

Many of the donated objects contribute to an atmosphere of mourning, which makes the museum resemble a memorial. In two instances mining-related objects are juxtaposed with references to the Second World War. The first example is "Coal Camp Exhibit: Typical Boone County Home Furnishings in the 1930's and 1940's," in which a copy of the *Charleston Gazette* with a headline announcing the bombing of Pearl Harbor lies on a small homemade-quilt-covered bed, along with another local paper and a copy of the *Saturday Evening Post* from the same era.[8]

In a vitrine across the room, a copy of a letter from a dying father and son, trapped by a mine explosion in Tennessee in 1902, is placed next to a newspaper with a headline marking the end of the Second World War: "Japs Bow to Terms." This quietly distressing display creates an affective link between the nationalist sacrifices of war and the sacrifices of coal miners and at the same time recenters coal miners in a civilizational/racial discourse as American national subjects. Although there is no historical rationale for the association of these particular items, in the past the link between

Pearl Harbor headline, Coal Heritage Museum, Madison, West Virginia, 2004.
Photograph by the author.

mining and war has been overt, as when miners were excused from the draft. Indeed, this slippage between mining and war is currently reflected in the national arena by the use of the iconic cry of the underground miner, "Fire in the hole!" in military operations and war games.

A lithograph displayed in the museum underlines these connections. The print, *The Price of Freedom,* shows an older white man and woman bowing their heads before a coffin draped with an American flag, surrounded by helmeted soldiers. With no overt reference to coal mining, this image reflects the felt articulations between mining and war, sacrifice and American national identity, as it wordlessly expresses the connections between masculine sacrifice in war and in the mines. As a repository of what many coalfield residents find valuable, the museum's collection entangles the patriotic grief evoked by the mourning parents in *The Price of Freedom* with the tragedy of miners' deaths underground. Through its emphasis of these linkages, the museum centers the coal heritage story on the *national*

Newspaper headline announcing end of World War II and mine disaster note, Coal Heritage Museum, Madison, West Virginia, 2004. Photograph by the author.

identity of the coalfields and articulates coal heritage through a discourse of masculine sacrifice.

The Beckley Exhibition Coal Mine is located off a major interstate highway through the coalfields. A state park has been constructed around the mine, with several original coal camp buildings from different locations that have been moved to the spot, in a re-creation of some of the features of a company town. These include a large, white, frame supervisor's house

Robert Prichard, The Price of Freedom. *The artist is from Madison, West Virginia.*

with a wraparound porch, a one-room schoolhouse, a church, and two examples of miners' dwellings (a one-room "shack" for a single miner and a
three-room company house for a miner with a family). The Youth Museum,
featuring a "Mountain Homestead," is located beside the church. The exhibit
depicts "a typical settlement on the Appalachian frontier" and consists of
a simulated homestead from the nineteenth century, including a barn, a
woodshed, a workshop, and other outbuildings in addition to a log house
filled with rustic antiques (Beckley Mine 2007).

The spatial association between the "Mountain Homestead," the Exhibition Coal Mine, and the re-created coal camp conforms to an unspoken
narrative of national development. The Youth Museum's Web site specifies that the homestead represents life in Appalachia in the late nineteenth
century, when Turner was lamenting the closing of the western frontier.
Naming this late nineteenth-century Appalachian homeplace a "settlement
on the Appalachian frontier" accomplishes several things. It reiterates the
logic of Appalachian marginalization; as a backward place, Appalachia
always represents America's past, therefore the Appalachian homeplace is
culturally anterior to its actual date in history. At the same time, it locates

the frontier as a quintessentially American place/process. The park's spatial association of the settlement (homestead) and the mine eternalizes and naturalizes an American frontierlike relationship to the land and natural resources, as embodied in the communities of geographic expansion and industrial resource extraction that are represented (Tsing 2005). These abstracted sites represent a traveling set of relationships between people and land that are essential to American national identity (Barnes 2010).

The gift shop includes an exhibit of old photos, domestic antiques, and old coal-mining equipment. The shop offers Appalachian arts and crafts and kitschy folk items such as replicas of old mountain cookbooks. The mine itself is an actual former mine that operated from around 1890 to 1910. Visitors board a small train that begins under a shelter and then winds its way into the mine. The tour guide during my visit was a white retired coal miner who clearly enjoyed recounting the harrowing conditions that pertained while the mine was being worked. He described how the miners, working only with hand tools and dynamite, drilled holes and loaded shot in a thirty-six-inch seam (i.e., the roof was thirty-six inches high) while lying on their sides or stomachs, then scurried out of their holes as fast as they could to avoid getting caught in the blast. Then they used shovels to load the coal onto cars.

The mine was well lit for the tour, but the guide momentarily turned off all the lights to show us how dark it could be underground. He explained all the equipment and answered questions from members of the small tour group about innovations in mining technology through the twentieth century, such as that of the continuous miner, and discussed developments in safety equipment like rock dusters, which coat the coal face with limestone dust to reduce the risk of explosions. He showed us some of the hazards of underground mining, such as the large round fossils called kettle bottoms (because of their shape) that can suddenly fall out of the ceiling without warning. He explained the purpose of chits and told how crooked coal operators would cheat miners by removing their chits or shortchanging them for their loads.

He also told a story about when the first women miners were hired in the mine where he used to work. He said that miners usually relieved themselves in the empty finished places in the mine, since the darkness provided a kind of privacy; it was so dark you could easily see if anyone was approaching by the light on their hat. But when "women came into the mines in 1972" the company was forced to provide portable toilets. The men miners scoffed and said, "If you're going to use them, you'd better clean them." But, he continued, "I came out of here [the mines] in '88 and I never seen one with the plastic [wrapping] off." The women had evidently decided not to use the facilities either. As the guide deadpanned, "Well, it's dark in the mines." We all laughed.

This "all cats are gray in the dark" tale was part of a tour narrative that emphasized the extreme working conditions that miners endure. It invited us to imagine conditions in which people (even women!) might have to relieve themselves in a dark hole underground. His lengthy focus on the mining techniques that were used when the exhibition mine was in operation provided a dramatic background for a subsequent story of progress in mining technology throughout the twentieth century. In this context, the presence of women in the masculine-coded mine indicates both the increasing liberalization of American society and the development of mining technology. Constructing mining as naturally masculine, this story erases the fact that around the world and throughout history women have worked and continue to work as miners, with or without mechanization (Hilson 2003; Moore 1996). This narrative naturalizes progress and elides the political struggles that have led to worker protection laws and affirmative action and that have spurred the development of mining technology.

This narrative of continual progress throughout the twentieth century provides a cultural context in which MTR becomes simply another step in an inevitable march of technological innovation. Viewed from the context of the lifeworld of the early twentieth-century underground miner, technological innovation in mining does doubtlessly benefit working people. But the popular narrative of capital-driven technological progress as represented in the mine tour is an overgeneralization of this benefit. Because technological innovation has improved people's lives in some cases, it cannot be assumed always to do so, or to do so unequivocally. This progress narrative forestalls consideration of the physical, social, and environmental costs of increased productivity and mechanization.

The exhibition coal mine is located in a historical park that replicates the past orientation of the Coal Heritage Museum. The mine supervisor's house is furnished with fancy antiques, and in addition to including representative bedrooms its upper story also contains a replica of a coal camp post office, doctor's office, and barbershop. The miners' dwellings are decorated with working-class antiques. Touring the houses, I was struck by the size of the furnishings, the stairs and hallways, and the rooms. The furniture was much smaller than furniture of today, including the table, chairs, sofa, and beds. Even in the supervisor's house the stairways and halls were quite narrow. I remarked on the small size of the beds to a fellow tourist in the miner's three-room house. She replied that the display was quite accurate to her memory, and that everything actually had increased in size since the time represented there, including people. This woman's perspective was one of a person who had once lived in a coal camp house like the one we were touring. Visitors to these sites are very often former coalfield residents or the children of coal miners who have formed a kind of coal diaspora as closing mines, mechanization, and increased productivity have decimated the ranks of people employed in mining. Ned, the county official involved with the creation of the Hatfield–McCoy Trails System, who was also active in the development of the Coal Heritage Museum, reported that most of the people interested in the museum and other heritage sites are "probably . . . related to people that's been in the industry. . . . You know Boone County [had] a vast population. Like I said, it's the leading coal-producing county, so a lot of people have left here, when the mines went down in the fifties, and they still want to come back and see what's happened in that span of time, and they get a lot of information down there."[9]

The existence of this ready-made audience, hungry for coalfield nostalgia, implies a "before" and "after" that is never exactly specified in these heritage sites. The era of coal heritage seems generally located in the 1930s through the 1950s, based on Ned's comments and some of the Second World War memorabilia, but the cutoff point is never clear.[10] For instance, the one-room schoolhouse in the park is divided into two sections. One represents a coal camp schoolroom, with antique desks, early twentieth-century textbooks, and a list of amusingly strict school rules. The other side of the building is largely empty, except for a shelf containing donated items from the community. The collection is a hodgepodge of anything old related to education, including a mimeograph machine and an overhead projector, which are certainly obsolete but seem not to match the degree of historical interest (yet) of the other things in the schoolhouse display.

This consistent past orientation combined with the lack of historical specificity allows "Coal Heritage" to stand for a homogeneous and shifting time of "coal culture," a nostalgic utopia/dystopia that is conceptually separated from the complicated reality and history of coal mining. It creates a sentimental story of coal company towns as spaces apart, where a simple work ethic and class structure determined everything completely, as opposed to the complex world of the modern coalfields, where even white women (must) work outside the home, the state is a bigger employer than the coal industry, and coal mining itself frequently looks more like construction or road work than the iconic image of the underground mine. The power of this conventional story of what coal means is reflected in the difficulty of including modern surface mining in popular representations of the coalfields in the national media.[11]

The Coal Heritage Museum and the Exhibition Coal Mine reiterate the commonsense understanding of the "coalfields" as a place defined by coal and glide over the politics of this designation through nostalgic allegories of the hardships of coal camp life.[12] These sites freely mix symbols of national identity with images of coal-mining sacrifice, telling a romantic story about Appalachia's place in America, as with the display of the headline announcing the attack on Pearl Harbor and its placement on the small, homemade-quilt-covered cot. The mine supervisor's house and the three-room coal camp house provide class-specific domestic contexts for the male-centered work life of the mine. In the three-room miner's house the kitchen table is set with colorful tin cups and plates, as if awaiting a father's arrival for dinner. These heritage sites offer representations of the national symbolic and affective domesticity that locate the practice of coal mining, and the history of the coal industry, in the context of a gendered and racialized American national identity (Berlant 1997).

Coal History

A thorough look at the history of coal mining in West Virginia creates a different sense of nostalgia. The horrendous conditions of early coal mining helped fuel a strong labor movement that persisted in its efforts to organize and protect the rights of miners over fifty years. In the pre-union era, striking miners and their families were frequently turned out of their homes because occupancy of the company houses depended on continued employment of the male head of household. Miners were often forced to sign "yellow dog" contracts, according to which they could lose their homes if they joined the union. In one of the first battles of the West Virginia Mine Wars, striking miners from Paint Creek and Cabin Creek in Kanawha County, which are today surrounded by an enormous spread of mountaintop mines, were thrown out of their company houses and forced to spend the winter in a tent city in the forest, where they were vulnerable to the weather and to attacks from the company's hired thugs.

Four years later, an army of union miners gathered to march on Logan and Mingo counties in an effort to install the union there by force. The pro-industry sheriff of Logan, Don Chafin, deputized the "better citizens" of Logan to defend the antiunion stronghold of Logan County, and the Battle of Blair Mountain ensued, one of the most violent episodes in American labor history (Corbin 1981; Shogan 2004). Stories of these battles evoke nostalgia for times of solidarity and mass action and still animate the labor movement in the form of union songs and quotes from Mother Jones.[13]

The activist network Friends of the Mountains has been working for years to have this battleground protected from MTR. In 1998, environmental activists sued the West Virginia Department of Environmental Protection for failing to enforce elements of the Clean Water Act requiring a "buffer zone" between mine activity and streams (Clines 1999a; Janofsky 1998). Their suit was upheld by the federal appeals court. In reaction, Arch Coal closed the Daltex mine on Blair Mountain, one of the largest MTR mining operations in history (Ward 1998). The coal industry and state political leaders claimed that this ruling was a threat to all mountaintop mines and by extension to coal mining in general because of its "rigorous" interpretation of environmental law (Clines 1999b, 1999c). Subsequently, the Bush administration weakened the Clean Water Act to permit mining in and around streams (Warrick 2004). Coal companies also started submitting smaller permit requests that attracted less attention but in aggregate achieved the same goals as enormous permits like the one stopped on Blair (although they are now turning to even larger mines). Activists reacted to these changes by changing their focus on Blair to historical preservation. They collected and mapped artifacts from the battle, mostly bullet casings, and they presented their case repeatedly to the West Virginia Archives and History Commission before the site was finally added to the National Register of Historic Places (Dao 2005). The site is currently delisted but if the listing is reinstated, the area will be protected by the National Park Service, making mining in the area more difficult (Ward 2009a, 2009d; Nyden 2010).

Many of the individuals involved in this ongoing effort have strong ties to Blair Mountain, including Bill, the activist fighting to preserve his family homeplace in Blair. His mother was a child during the battle; she remembered her family hiding union men in the barn and the creek outside the house running red with blood. Shawn, who for nearly twenty years has done much of the legwork and archival research trying to establish the historical value of the area, lives at the base of Blair Mountain, in Logan County, and has family members who fought in the battle. For these men, the place, the physical site of the battle, is an essential piece of the area's history that must be preserved to honor the memory of the miners who fought for the union. They refuse to sacrifice the land to the corporate or national interest; rather, they insist on the right of coalfield residents to determine their future and on the inherent value of their communities. For instance, Bill argues that mountain communities' reliance on water wells is not an indication of historical backwardness or Third World–style poverty; rather, it represents an alternative and equally valid way of life. He contests the articulation of mining and progress when he exclaims sarcastically that sinking creeks and ruining wells, then requiring people to take city water, is economic development ("That's improvement!"). Shortly after the environmental lawsuit and the closure of the Daltex mine, the Blair Mountain Historical Organization held a reenactment of the "Miners' March" from Marmet to Blair that preceded the Battle of Blair Mountain. From the marchers' perspective, they were honoring coal miners and their heritage, and indeed the Heritage Museum does house some bullets from the battle. But as a "symbolic transaction" (Coombes 2003, 14), the commemoration failed to create new constituencies around coal heritage. The marchers were pelted with eggs and physically assaulted by laid-off MTR miners holding a counterdemonstration. Several march participants were physically attacked. Bill, who was seen as one of the instigators of the lawsuit, was violently assaulted by someone he had known since childhood. In the following summer of 2000, when I went to Blair for the first time, I noticed that someone had spray painted the concrete railing of a bridge leading into the town of Blair with the words "Fear God Bill——."

This violent reaction to the mine closure, and to the Blair Historical Organization's activities, which the mountaintop miners considered to be more about stopping MTR than about honoring the union, signals the ever present intensity of emotion around mining in the coalfields. This symbolic transaction may have faltered around the definition of locality; is the community embodied in a place or is its existence tied to its economic function? This distinction reduces to two incompatible versions of the place; where the industry sees only labor or obstacles to production, the Save

Blair Mountain movement claims that the communities have intrinsic worth, regardless of their economic role.

Newspaper articles about these conflicts frequently refer to the issue as a new battle for Blair (Dao 2005; Janofsky 1998). Rather than a new battle, this fight is a continuation of the battle, begun in the early twentieth-century Mine Wars, about who controls the place and how the community will be defined. The sides of the battle have been reconfigured, however, as the union withdrew its initial support for having the battle site named to the National Register of Historic Places.[14] These shifting sides indicate the unbounded, situational spatial politics of defining universal and particular. As during the Mine Wars, the fight today is over what place coalfield communities have in the American nation—what kind of rights they have in their homes and property, what level of citizenship they are granted. For the Save Blair Mountain movement, this history of struggle is also part of coal heritage. Whether as workers or as stakeholders in minerally rich land, the ability of coalfield residents to control the fate of their physical environment as full citizens of a democracy has always been problematic in a nation that depends on the exploitation of their labor and the land.

The National Register of Historic Places and National Heritage Areas

Another way to approach these stories is to take a look at the federal institutions that support the different projects. The National Register of Historic Places (NRHP) is under the control of the National Park Service of the Federal Department of the Interior, which has a say over what can be done with the places listed there. The register's Web site states that federal intervention in property owners' business is unlikely, with one important exception: "consideration of historic values in the decision to issue a surface mining permit where coal is located in accordance with the Surface Mining Control Act of 1977" (NRHP 2006).

Predictably, the companies view this potential interference as an outrage. The *New York Times* quoted the vice president of Dingess-Rum Properties, Inc., the land corporation that leases the land to the coal companies: "We are going to resist vigorously any attempts to take away our property rights. We have a right to exercise our lawful and legal right to mine coal, remove timber and drill oil and gas wells on our property." The activists are equally clear about what the listing means: "This is a breakthrough nomination in terms of taking a chink out of the power of coal companies to dominate our land" (Dao 2005). In this case, federal oversight of the land actually increases the self-determination of the residents. According to Shawn, the final decision on whether the place is listed is made by a vote of all the landowners affected, whether their property is thousands of acres or only half an acre, effectively negating the overwhelming dominance of coal.[15]

If the NRHP entails this small but significant degree of federal oversight on private property, the same cannot be said for the National Heritage Areas (NHA). In 1996, eleven counties in the southern West Virginia coalfields were designated the National Coal Heritage Area (NCHA). The goals and mission of the NHAs are quite different from the mission of the NRHP; the emphasis in the heritage areas is on local control. As an area that encompasses eleven counties, the NCHA is clearly a much more ambitious enterprise than the effort to have the Blair battle site listed on the NRHP. The NCHA Web site explains the reasons for such an enterprise:

> National Heritage Area designation recognizes the coalfields of southern West Virginia as a treasure of national importance. This designation provides an opportunity to tell the story of the southern West Virginia coal mines to visitors from outside the region as well as local residents. Because it is a project that stresses partnerships, land ownership and land use decisions remain at the local level. At the same time, southern West Virginia will gain new prominence on national tourist maps. Because it is a project for which the National Park Service provides technical assistance, the region is included in materials used to describe sites around the country—sites important to the history of our nation. As a plan takes shape and is implemented, the heritage area will become a much-needed focal point for the revitalization of southern West Virginia, encouraging residents to embrace their heritage with a new sense of pride and ownership. (NCHA 2006)

This statement includes several of the most important things to know about the NCHA. First, it is a campaign for the representation of West Virginia on a national level. Second, and following from the first point, the NCHA is part of an effort to revitalize southern West Virginia and allow residents to feel pride in their heritage. By constructing the history of coal mining as "national heritage," coal heritage centers West Virginia in American national history and naturalizes its status as a sacrifice zone. Significantly, heritage preservation relies on private property (Bommes and Wright 1982, 273). Thus, coal heritage is part of a series of partnership initiatives, like the Hatfield–McCoy Trail System, that at least partly eschew federal or state government authority in favor of private control. Finally, the statement's emphasis on local control of the NCHA reveals part of the irony and subtle (or not-so-subtle) pro-industry agenda of the heritage movement. In an area where more than two-thirds of the land is owned by corporations, what does it mean to say "land ownership and land use decisions remain at the local level"?

The Strategic Management Action Plan (SMAP) (Parsons, Brinckerhoff, Quade & Douglas 2001) of the NCHA reveals more about the contradictory goals of the heritage movement as it suggests ways to implement them. It includes repeated references to nationally held stereotypes about West Virginia and the coal industry and the necessity of an opportunity to more accurately represent the region and the industry in the national arena. It begins, "The [NCHA] encompasses hundreds of square miles of rugged industrial landscape where hard-working miners of all races labored to extract and transport the coal which shaped modern America." It continues, "The coalfield story is a uniquely American story which is expressed in the customs, communities and stories of this remarkable region" (i). However, the authors see stereotypes as an obstacle in the way of the NCHA's goal of attracting tourists to the area: "The coal heritage story, while important and compelling, has marketing challenges related to regional stereotypes and also to stereotypes about mining that provoke negative images"(42).

At the heart of the NCHA management plan is the hope that tourism will become an important industry in southern West Virginia. The SMAP puts it this way:

> Unlike "rust-belt" industrial districts and the downtown cores of older industrial cities, the coalfields have never benefited from a countermovement [to deindustrialization and population loss]— "gentrification"—that has converted obsolete structures and districts to the uses of service- and information-oriented industries and of consumers who search for touches of "authenticity" amid the standardized offerings of a global economy. Tourism is one of the service industries that gentrification has served, and the National

Coal Heritage Area represents an attempt to capture some of its
benefits for southern West Virginia. (9)

This fervent hope to jump on the postindustrial/postmodern band-
wagon is unfortunately mired in the realities of coalfield geography, however, because for one thing, the area is not yet industrially obsolete—coal
production is as high as ever. The presence of the active coal industry in
the heritage area is problematic for the NCHA's plans, as is the negative
heritage of coal mining. The SMAP laments, "Coal extraction, processing,
and transportation has left [sic] visible impacts on the landscape. Population loss, depressed economic conditions and uneven waste disposal practices have compounded negative visual impacts. As a result, blight and
clutter is [sic] commonplace" (95).

The SMAP includes plans for overcoming these obstacles in order to
take advantage of the "compelling" story of coal heritage. It argues that if
West Virginians build it, tourists will eventually come. "Like all heritage
areas celebrating regions heretofore unappreciated as tourism destinations,
the National Coal Heritage Area faces several marketing challenges. . . .
Conveying the complexity, diversity, and relevance of the coal heritage story
to visitors from throughout the US and Canada may require overcoming
preconceived notions about the industry and West Virginia" (75).[16] This
statement expresses some of the contradictions of the heritage movement
and the difficulties of developing a coalfield tourist economy. As a marginalized region, the area is stigmatized in the national media, popular culture, and scholarship, as well as bearing the physical and environmental
effects of serving as an interior resource colony, as the SMAP notes. But at
the same time, the processes of this economic and cultural marginalization
are generative of "coal heritage" and the very regional identity that the
NCHA celebrates. This contradiction is present throughout the SMAP. The
authors frequently acknowledge that tourists are reluctant to plan vacations to West Virginia because of "Appalachian stereotypes" that "perpet-
uate distorted and inaccurate perceptions of the region and its people"
(60). However, according to the SMAP, those enterprises that succeed in
attracting tourists do so on the basis of these stereotypes. For instance,
when praising the successes of the coalfield ATV trail system, the authors
state that the "Hatfield–McCoy Trail System . . . enjoys one of the few im-
primaturs [sic] recognized by tourists and associated with West Virginia.
The famous feud elicits curiosity even today" (74).

Nonetheless, overcoming cultural and economic marginalization is a constant theme in the NCHA Strategic Management Action Plan. The SMAP illustrates the major goals of the heritage movement, many of which hinge on reimagining the region in the national arena as an attractive and safe place to visit. As the authors state, "The NCHA provides the opportunity for the region's story in all of its dramatic and complex aspects to be told in an authentic and historically accurate manner" (60). The authors repeatedly claim core national status for the coalfields: "Eventually the NCHA may be able to take advantage of international visitors, who now show interest in getting away from their traditional destinations (major cities, national parks, and Florida) and experiencing more of the real America" (64).

National Heritage Areas and Mountaintop Removal

The NCHA Strategic Management Action Plan's claim that the NCHA is the real America needs to be backed up by successfully conforming to certain hegemonic American standards. One key to achieving the NCHA's goal of becoming a tourist destination, as recommended by the SMAP, is to improve local infrastructure. "A great deal of the viability of the NCHA, particularly interpretation opportunities and visitor services in core counties, depends on improvements to the transportation system" (Parsons, Brinckerhoff, Quade & Douglas 2001, 59). Local roads are noted to be twisty, steep, narrow, and dangerous. One such road, Route 10 through Logan County, is by local reports one of the most dangerous roads in the nation. But the danger is not inherent in the road; it is the presence of overweight and speeding coal trucks that make it deadly (OHVEC 2007). Drivers are paid according to the tonnage they deliver instead of by the hour, which encourages extremely hazardous conditions.

Two huge road projects in southern West Virginia have recently attempted to take advantage of the coal industry in the ways that its supporters have repeatedly suggested (R. Scott 2001). The King Coal Highway and the Coalfields Expressway involve allowing companies to mine multiple seams of coal (according to activists, without a proper permit) and then prepare the postmining surface for road building (Chafin 2005).

King Coal Highway/I 73-74 (Exhibit E).
In cooperation with the Department of Highways and the
Department of Environmental Protection, the coal industry

plans to construct (to rough grade) 5 miles of the new King Coal
Highway/I 73-74, with 2 connectors . . . saving the taxpayers an
estimated $90 million dollars. (U.S. Senate 2002, 23)

The King Coal Highway will connect Williamson, in Mingo County, to
Welch, in McDowell County, and the Coalfields Expressway will connect
Beckley, in Raleigh County, to Welch. McDowell County is currently one
of the most economically depressed places in the United States. During
the 1930s, Welch was a boomtown. The neighboring town of Keystone,
where my mother was born, was a notoriously thriving place with gambling
and red-light districts. These roads are a key part of a project to revitalize
the economy of McDowell County, with an industrial park and a new federal prison planned near their interchange. But activists argue that these
roads are primarily being constructed for the convenience of the coal industry, allowing access to remote areas for mountaintop mining and making
the transportation of coal easier.

City hall, police department, and coal processing plant, Keystone, West Virginia, 2004.
Photograph by the author.

For the coal heritage movement, these roads are part of a vision of the future in which reimagined coalfields will become an international tourist attraction and downtrodden coalfield communities will become "visitor experience zones." Relics of the company town days (tipples, wooden company stores, etc.) are important elements of this visitor experience, and the SMAP urges their protection: "Valuable resources are being lost to demolition, neglect and vandalism. Each year more irreplaceable resources are gone forever. In partnership with the State Historic Preservation Office, regular updates on the inventory and evaluation of unique and endangered structures should be conducted" (Parsons, Brinckerhoff, Quade & Douglas 2001, 92). One of the most valuable resources of the NCHA is, in fact, Blair Mountain. The SMAP states, "The site of the 1921 Battle of Blair Mountain [is] the only site in the region deemed to be of national significance according to established preservation criteria" (50). Despite the danger the NCHA poses to this valuable site, MTR cannot become an issue. Rather, the authors tap-dance around the elephant in the room:

Although West Virginia's economy has shown improvement during the recent national boom, no area in the state has had more difficulty rebounding from de-industrialization. The southern West Virginia economy still depends heavily on coal mining, but the increased productivity of the mining industry has led to heavy job and population loss in NCHA counties. Against this backdrop, there continues to be conflict over mountaintop removal mining practices. One outspoken McDowell county native has been speaking out against the mountaintop removal practices of today's coal mining companies while writing novels celebrating the cultural history of mining.[17] From another perspective, mining interests claim that mountaintop removal is needed, both to keep the coal mining industry viable in West Virginia, and to produce more flat land for development. *The NCHA cannot and should not become mired in the politics of mountaintop removal.* However, if it is to be a successful progenitor of the post-coal economy of southern West Virginia, the NCHA must be a strong advocate for preserving and interpreting the physical and cultural remains of the coal boom era. (59, my emphasis)

The apparent contradiction of preserving coal heritage sites while not opposing MTR is addressed at another point in the SMAP. The authors mention the necessity of moral considerations, namely, respect for "local ownership of the process and products" of coal heritage tourism, and they cite a National Park Service study that concluded, " It is the stories and culture of this area, rather than the sites, which are most important." The SMAP goes on to conclude, "Public Law 100–699 [which established the NCHA] seems to have recognized this in giving 'cultural values' equal standing with historic and natural resources in its directives to the Park Service" (55).

From an institutional perspective, then, the NCHA's Strategic Management Action Plan is concerned with "local" control but not with questioning the status quo. The authors want to conserve coal heritage resources but in the end emphasize the relative importance of "cultural values" over actual sites that may be destroyed through the continued activity of the coal industry. This reflects the divergence of the NCHA, which is an economic development initiative, from the desire of coalfield residents to remember their history and preserve their communities. Meaning, emotions, and cultural memory are rooted in objects and places (Stallybrass 1999; Sturken 2004). But individual and community memories are simply another resource for the coal heritage movement in its economic development form. The SMAP's concern with culture and local control is ironically antidemocratic and tied to the interests of coal boosters who see MTR as a way to modernize and rationalize the coalfield economy (U.S. Senate 2002).

Almost any element of coal-mining life can be stripped of its politics and become fodder for commodification into coal heritage. In a section titled "Journeys through the Region and Interpretive Themes," the SMAP focuses on the following "interpretive themes": the coal business, working in coal, the company town, mining technology, and crisis and renewal. Most striking is the discussion of the company town. The authors write: "While [the company towns] were captive communities, in the sense that residents lacked the privacy, rights and diverse job opportunities that workers in urban settings enjoyed, they also offered amenities, such as electricity and a broad array of consumer goods that were hard to come by in rural West Virginia before the 1930s" (Parsons, Brinckerhoff, Quade & Douglas 2001, 7). Likewise, according to the SMAP, the company stores were not all bad. Yes, they charged higher prices than other stores (because of logistics, not greed, according to the authors), but they also provided "wiser choices and easier credit" (7) than were available in other rural places. This is a nice spin on no selection and debt peonage.

The SMAP echoes a frequently cited advantage of the company town, its cultural and racial diversity. The section on company towns ends on this positive note:

> While native white, African American and immigrant ethnic groups were generally housed separately . . . the opportunities for interaction . . . and the unsegregated character of underground working conditions contributed to a richer cultural diversity than any of the three groups had encountered elsewhere. Even shared conditions of hardship and danger in the mines contributed to a sense of community solidarity that residents often later recalled fondly. (7)

This passage simplifies the history of segregation, eliding the role of race in the struggle for unionization and the hierarchical job segregation that structured race-based economic inequalities in the coal camps (see Eller 1982; Trotter 1990).

Regarding the struggles for unionization, the authors comment, "Stories of these battles and how they helped shape history and the union movement could be told [in visitor experience zones] through recorded interviews, exhibits, and reenactments" (Parsons, Brinckerhoff, Quade & Douglas 2001, 30). But this is not the only topic of potential interest for tourists:

> Another important, although less dramatic, battle, was the effort to protect miners as they worked through stronger mine safety and health legislation. The depiction of this effort could be reinforced through linkages with the mine disaster database now being developed by the National Mine Health and Safety Academy in Beckley, and the National Coal Miners' Memorial being planned in Nellis. (30)

Unfortunately, this element of coal heritage has also slipped loose from its temporal categorization. Recent mine disasters have revealed that the battle for miners' safety is far from over. Nothing better shows the emptiness of the concept of coal heritage than this attempt to present what is actually an ongoing struggle for workplace safety in an antilabor atmosphere as a past event that can be educationally and entertainingly viewed by tourists.

The NCHA Strategic Management Action Plan represents an uncomfortable compromise between local economic boosters, who envision and hope for a postcoal economy based on tourism, and the imperative of rapid, continuing extraction of coal by the coal industry. Whatever economic development can be generated by tourism must happen in the interstitial spaces of coal production as required by the industry, which still holds the economy hostage by virtue of owning two-thirds of the land and having underdeveloped the region over the course of more than a century of dominance. At the same time, the coal industry's continued operations in the area are destroying the natural features that have historically been West Virginia's most successful tourist draw, and, most undemocratically, are permanently altering the future of the remaining coalfield communities, postcoal.

"Coal heritage," as told by the coal heritage movement (if it has any substantive meaning at all) refers to a historical period when working people in coalfield communities had a small share in the industry's profits. Those coalfield residents invested in preserving the material traces of their communities, whether in the form of collected artifacts or in the landscape itself, are trying to hold on to distinctive ways of life that happened, and are still happening, in a particular place. But the NCHA's investment is in having the region reimagined as a safe, average part of modern America— part of the homogenized global economy, with some remnants of an interesting and "authentic" past to engage tourists. Despite the ongoing extraction of coal, the holders of this vision want West Virginia included with the rest of America in the so-called postindustrial era. This is a vision of a region that would offer for tourists plenty of safe folksy chain restaurants staffed by locals fluent in interpreting coal culture and colorful Appalachian stereotypes such as the Hatfield–McCoy feud. Hillbilly jokes and images are welcome on restaurant placemats, as long as the "blight and clutter" associated with them have been cleaned up outside. In the rationalized economy envisioned by the coal heritage movement and other coal industry supporters, these bits of local specificity are to be contained in commodifiable locations, allowing the area in general to be safely Americanized to attract the curious but fearful tourists whose sensibilities may be offended by "negative images." This is a vision of a postmodern economy to be sure, selling a commodified version of the coal cultures that economic exploitation has generated to visitors who must be protected from the real effects of historical and ongoing coal production.

Appalachia is a place caught in contradictions of geographic scale; it's an "all-American" region with Third World problems. For the last century, it has been a national sacrifice zone where the coal industry has pushed the limits of the possible in its efforts to feed the national appetite for energy and meet its own profit imperative by extracting coal. The bloody, messy process of this natural resource extraction has involved the sacrifice of countless workers' lives and health and, increasingly, of the landscape itself. The two accounts of history that I have presented here illustrate how the past constructs and inhabits the future. For the Save Blair Mountain movement, stories of past labor activism offer hope for the salvation of the landscape and the mountain communities it has sheltered. The coal heritage movement, however, blends Appalachian history into the national through discourses of sacrifice and development. In this account, the patriotic sacrifices of miners constitute the terms for Appalachia's membership in the American nation. As an economic development project, however, coal heritage ironically commodifies local memory into a new resource for extraction.

CHAPTER 6

Traces of History:
"White" People, Black Coal

Appalachia: A Problematic White Space

Appalachian identity is tricky. This trickiness has something to do with the historical permutations of racial formations in the United States, as well as relating to the complex intersections of these with gender and class formations and the ways that people relate to nature. These intersections and cultural articulations help explain both the flexibility and the intransigence of social categories. Race, gender, and class, along with other cultural formations like region and the human/nature relation, interact in ways that are mutually reinforcing yet overdetermined in their possibilities. These complications enmesh Appalachian whiteness in a number of contradictory stories that both complement and trouble the racial project of whiteness (Omi and Winant 1994). In one, Appalachian whites are the most "pure" in America. According to this story, white hillbillies are descended from unadulterated Scotch-Irish stock and are relatively innocent of the troubled race relations of the South or the northern cities (Inscoe 1999). In this way, they are articulated with idealized notions of America's past and with the pious and devoted nationalism of the heartland. In another story, Appalachian whiteness is often defined by references to Native America, sometimes via claims of Native ancestry, sometimes through the trope of the frontier, as epitomized in a proliferation of place-names and legends. And in yet another contradiction, hillbillies are portrayed as degenerate whites: inbred, violent, and backward. Traditionalism here represents irrationality and brutality, including violent racism, deviant sexuality, and oppressive gender relations.

The story of degeneracy resonates strongly with the American national narrative of differential worthiness, as it accounts for coalfield poverty in terms of a pathological hillbilly or "white trash" culture. The poverty and disempowerment of the coalfields belie the dominant American cultural narrative of the rewards of productivity, as well as the idealization of "white people,"[1] and these contradictions must be accounted for. For coalfield activists, coal dust is a sign of the oppressive environmental, economic, and social conditions of coal mining. This black dust signifies the association between the industrial process of mining coal and the poverty, hardship, and suffering of the coal camps. But the symbolic association of coal mining and blackness carries other connotations as well. In its currently hegemonic form, whiteness is articulated with things like property, progress, and hygiene. Coal mining, poverty, and underdevelopment undercut this story. In the coalfields, as elsewhere, whiteness is haunted by the specter of "white trash"; failures of ideal whiteness are explained by metaphors of racial Others and the places they live.

MTR happens in a "doubly occupied place" (K. Stewart 1996); Appalachia is economically occupied by the coal industry, by its marginalization in the national culture, and at the same time occupied by the people who live there, who must find ways to negotiate their lives within these other forms of occupation. This chapter examines one aspect of this double occupation, the negotiation of white American identity in the coalfields. In the national imagination, Appalachia is a "white" space, and hillbillies, whether idealized or demonized, are imagined as white people. This reflects the centrality of racialization to American national culture. It is not my intention to determine the racial attitudes of coalfield residents. Rather, the point is that race and racial distinctions are fundamental symbolic categories in American culture, and the logic of differential worthiness that they reflect is deeply ingrained in other, seemingly unrelated areas, such as how we understand the economy and the human relationship to nature. Because we live in a social formation shaped by inequalities of gender, race, and class, these categories frequently inform our ways of interpreting the world, even when the concepts themselves are not part of a group or individual's discursive repertoire (Johnson et al. 1982, 252; Bettie 2003). The comments I analyze in this chapter are not interesting because they reveal some "truth" about particular individuals' racial attitudes; in fact they do not. Their interest lies in the way that these oblique references to race make visible some of the articulations between divergent cultural categories.

Whiteness is frequently experienced as a nonidentity by whites because of its role as the unmarked center of American racial formations (Dyer 1997; Frankenberg 1993; Perry 2002; Twine 1997). As the unmarked center, part of whiteness's power comes from its invisibility and its claims to universality (Goldberg 1993). However, there are specific valorized qualities associated with white identity, involved in American cultural citizenship, which come to light in the context of MTR. MTR is an extreme example of how the human relationship to land is structured by cultural formations like race and gender. The cultural politics of MTR reinscribe the logic of differential worthiness in a context that is nominally race-free. At the same time as Otherness is reaffirmed, white coalfield residents desirous of unmarked American citizenship are preoccupied with whiteness.

A Word on Whiteness and U.S. Citizenship

Whiteness as a racial project is related to American cultural constructions of citizenship, individualism, private property, and modernity, as well as through the mutually generative interconnections between these concepts. U.S. citizenship was explicitly racialized in one fashion or another until 1952. The Dred Scott decision of 1857 declared that black people were not and could not be citizens. From 1790 until 1870, naturalized citizenship was limited to "free white persons." After the Civil War, this was expanded to include "persons of African descent." This allowed black people a limited degree of citizenship, but the citizenship of Native Americans remained in question until 1940. People of Asian descent were not allowed to become citizens and were even barred from immigration, starting with the Chinese Exclusion Act of 1882. Ian Haney-López documents how the courts struggled with defining whiteness in immigration cases from this period. People desiring naturalized American citizenship, or trying to maintain their extant American citizenship were it challenged, had to argue that they were white. (Haney-López notes that in these cases no one claimed U.S. citizenship on the basis of being black.) Scientific accounts of the "Caucasian race" (which included some groups of Asian descent) were often at odds with "commonsense" ideas about who could be white. Again and again, the courts sided with "common sense," basing their decisions on a popular conflation of color, degree of "civilization," and fitness for citizenship. These court decisions, in turn, reaffirmed popular constructions of American citizenship as "white." The historically white face of American citizenship is visible in the exclusive focus of recent anti-immigrant movements on immigrants of color (Haney-López 1996).

Part of recent anti-immigration fervor has concerned the drain that undocumented immigrants allegedly place on state resources. The "illegals" that anti-immigration activists are worried about are poor and frequently rely on the social safety net (such as it is) for nutrition and health care. This is a commonsense articulation between poverty and illegality that seems to go without saying, but noting it underscores the connections between whiteness, American cultural citizenship, and private property. Historically, property ownership has been as intrinsic to American citizenship as whiteness. In the liberal political tradition, property ownership has been constructed as essential for an individual to have the capacity for self-governance (Goldberg 2002).[2] Through the limitation of full citizenship rights to white people, property ownership became a characteristic of whiteness. Along with not being able to vote or use the court system, black people were not allowed to own property in colonial America—indeed they were legally defined *as* property (Harris 1993). Even free black people in the eighteenth century had no legal protection for their enterprises or belongings (Kazanjian 2003, 63–64). The citizenship rights of black Americans were again curtailed in the post-Reconstruction period through discriminatory voting laws and segregation. Superficially race-neutral laws requiring literacy to vote, or enacting poll taxes, were aimed at black people; white people were excused via "grandfather clauses"—if your grandfather could vote, so could you. These and other measures derailed the promises of Reconstruction, leaving southern property relations much the same as they were under slavery. In the early twentieth century, white Americans used lynching terror as a method of maintaining their civic and economic privilege (Ware 1992, 179–80). This form of economic and political exclusion was repeated in California's Alien Land Act of 1913, which forbade "aliens not eligible for citizenship" from owning agricultural land, and through the dispossession and internment of Japanese Americans during the Second World War (Hong 1999). U.S. property law directly favored liberal British individualism over other relationships to land, denying Native Americans' rights to land, for instance, because their relationship to that land was communal, not individualistic (LaDuke 2005, 118; Harris 1993). This helped entrench the articulation of whiteness and individual private property.

The connection between race, citizenship, and property relations has been constructed and reproduced through the denial of citizenship rights to people of color: initially through slavery, later through racial violence, by means of the Chinese Exclusion Act, through the racially differential benefits of the New Deal, through anti–affirmative action laws, or through efforts to deny a social safety net to undocumented workers. These practices have created and continue to sustain a pool of cheap, vulnerable labor that benefits the middle class and the wealthy (Hong 1999; Mink 1990; Valocchi 1994). The historical economic privilege of whites has placed most American wealth and property in white hands. This white economic privilege has become its own kind of property (Harris 1993; Lipsitz 1998). Differential standards of police protection, disproportionate incarceration of people of color, periodic white racial violence, residential segregation, and the internment of Americans of Japanese descent during the Second World War, have further established property and citizenship as "white." State protection, in the form of laws and their enforcement, is essential for the existence of private property. Not only are people of color often offered no state protection from violence against their persons or their property, they have historically experienced dispossession and violence at the hands of the state (A. Gordon 1999; Hong 1999; Hurtado 1996; Ignatiev 1995; Kazanjian 2003).

Liberal citizenship theory understands society to be made up of individuals, each pursuing his interest in the market. According to Lockean labor theory, property is generated when individuals mix their labor with nature to transform and "improve" it (Cronon 1983; Curtin 2005; Shiva 2005). Private property is thus articulated with the modern ideals of development and progress. This societal origin story also implies that wealth follows merit—those with wealth have it because they are more industrious than others. But the industriousness of the wealthy also promotes the public good; with each individual pursuing his own interest in civil society, the free market will supposedly find the optimal use for every resource, including every piece of land (Blomley 2002). Thus, the market-based development of privately held natural resources is understood as progress, which helps construct other ways of using land as backward and inefficient.

By defining our own so-called free-market economy as the pinnacle of human development and freedom, American culture constantly judges its Others according to how they measure up. Property, whiteness, modernity, and development are articulated in a feedback loop of differential worthiness that continually reaffirms the privilege of the wealthy First World over the impoverished Third. This ideology is apparent in common idioms that suggest the "better citizens" of a place, or the "best people," are those who own property.

Theorizing Whiteness and Mountaintop Removal

To paraphrase James Baldwin, as long as people think they are white, there's no hope for the environment. Whiteness is a key element of the cultural context in which MTR makes sense. In the context of Appalachian cultural marginalization, the quasi-racial connotations of coal mining exert a particular discipline on white coalfield residents. The dichotomized Appalachian in national culture finds internal expression in white coalfield residents' desire to identify as ideal rural citizens, not backwoods "trash." Appalachia is a place on the margins of whiteness, where the fictions that maintain white identity and the racially inflected ideals of American culture tend to fail. The most vivid example of this failure is in property relations, which are a linchpin in the construction of white identity and American political culture. But whiteness is also problematized and performed in the micropractices of individuals who identify as white or who value concepts articulated with whiteness. Historically, the practices of everyday life were explicitly defined in racial terms, and through these practices individuals continue to make important claims about their identity and moral standing. In the coalfields, the cultural complex of coal mining, in particular underground mining, problematizes this construction of whiteness because of its association with suffering, poverty, and abjection. This specter of failed whiteness disciplines those coalfield inhabitants who strongly desire to belong to modern America. Ironically, MTR, by annihilating the place, promises such belonging.

Coalfield Property Relations

Marginal Property

Property ownership, a set of rights in regard to a piece of land or other object, is a legal construction; nonetheless, private property is a sacrosanct ideal in American culture. Private property is one of the cornerstones of the entire social, cultural, and economic system. Part of the sacredness of private property is related to its importance in defining American citizenship; ideally, America is a nation of self-reliant individuals. But natural disasters, environmental devastation, and the everyday basic interconnectedness of life constantly illustrate the fiction of individualistic private property. In the coalfields, property rights are stacked in favor of the corporation; the right of an individual to protect his or her little piece of land is tenuous, downright illusory, if the coal company that owns the hills above decides to alter the topography. An ironic illustration of this practical fragility of rights to property in land is what happened to the Logan County office of the West Virginia Department of Environmental Protection (DEP). The agency had to move its offices to a new location a few years ago because the former site, in the words of a Logan County resident, "started getting flooded out." What happened to the DEP office happens to individual households many times over; as the topography is altered by logging and MTR, new floodplains are created. Whole communities suddenly find their "real estate" fundamentally transformed.

As American citizens, homeowners usually expect to embody a certain way of life, a developmental and civilizational standard. However, because of the fictional nature of independent property, it is possible to own so little that your property works against you in terms of civilizational standards. "Postage-stamp sized" lots in former coal camps, and coalfield property in general, which exists in the shadow of corporate domination, work against the desires of the residents to live up to middle-class American standards. Steve, the DEP officer whose views on the local habits of land use were discussed in chapter 4, described for me the "private" land holdings (that is, noncorporate) in former coal camps that are often unincorporated with limited access to public services.

> A lot of these companies, as time has transitioned . . . sold [the former company houses] to the individual, you know, let them be responsible, let them pay property taxes . . . [and today] . . . you've got a little community, a house with a little yard, with the chain link fence . . . and it's all private property, and all their sewage goes straight into the river. Why? Because no one has room for septic systems. . . . I mean, it's a legal snafu. A lot of them would put them in, if they had a choice, but they can't get a permit . . . because they can't physically do it.

This story shows how much corporate land ownership is part of the taken-for-granted landscape of the coalfields. The corporate ownership of most of the land is so unquestioned that in our conversation Steve doesn't even think of it as private property—it has taken on a quasi-official status. The unsanitary conditions of these former coal camp houses are directly related to the company's miserly unwillingness to sell more than the minimal amount of property necessary to externalize the costs of the residences to their individual owners, but it has become naturalized as part of the character of the people. This illustrates how liberal citizenship, embodied in the property owner, disguises the interconnectedness of life: legal property creates a fiction, but not the reality, of self-contained units of existence.

Civilizational Dreams

Over and over again in interviews I heard people argue that MTR supplies something southern West Virginia lacks—flat land for development. "We don't have any bottom land," Sam told me. "We don't have any flat land here," Jerry repeated. I heard this argument repeated so often that it began to take on a kind of doctrinal reality. Thus I found my reality jarred when Shawn described how Arch Coal had dismantled the town of Blair: "We had two schools; the schools have been removed. Now there's a sediment pond there. Blair had stores and takeout restaurants. Bottom land? They say we need bottom land. I'd say we have plenty."[3]

These divergent accounts—the hegemonic argument that southern West Virginia is poor because there is simply no room in the steep mountains for economic growth to occur, and the view expressed by Shawn, that there is actually plenty of flat land already—reflect the way that property ownership influences how we read the landscape. At the level of the community, the long domination of corporate land ownership has led to the perception that the land is naturally unavailable for other uses. Shawn's ability to see through this pervasive corporate presence has to do with his intellectual skepticism about the coal industry in general, but for coal industry supporters this lack of developable land is so naturalized that MTR seems to offer a miraculous technological solution for the area.

If other kinds of development are impossible, it follows that the coalfields can be nothing more than a sacrifice zone. This explains why the Appalachian Mountains are being destroyed. MTR is a clear example of a place being sacrificed for the national interest. What underground mining does below the surface, and to miners' bodies, MTR is doing out in the open, to the very place, consummating and obliterating the coalfields. As a sacrifice zone, the region is forcibly Othered. MTR is not something that could happen in a "normal," unmarked, American place. It only can happen in an interior colony like Appalachia, which sacrifices for America but is not of America (Kuletz 1998; Lewis et al. 1978).

But wait—at the same time, MTR is a very American thing after all. It is a larger-than-life expression of the ideal of improving nature; according to some, it's not that different from building roads or malls. As some local industry supporters put it, "God put the coal here" for "us" to take and use, in a kind of Manifest Destiny; it would be derelict of us as Americans not to use the coal that God in his wisdom put here in the mountains to make our economy strong. And MTR allows more coal to be accessible than ever before, now that all the "easy" coal is gone; technological innovation is required to get the coal that God put here for us. Not only that, but MTR offers a way to overcome the natural obstacles to development that have impoverished the coalfields! Thanks to the flat land it provides, all kinds of new industries are possible. Several coalfield residents exclaimed to me how beautiful the reclaimed flattops were. "You'd think you were in Kansas," said one. Another thought it looked like Wyoming. These emotional reactions suggest that if we are not in West Virginia perhaps we are just in America. According to these coalfield supporters, MTR, for all its destructiveness, offers a way to Americanize West Virginia. For the residents, this would mean (in theory) a diversified economy, tourism, better infrastructure, a better reputation nationally, more commercial consumption opportunities locally—in short, more modernity.

In the eyes of industry supporters, the flat land created by MTR epitomizes the civilizational standard of property. Domesticated by corporate ownership, this flat land is an affirmation that nature is always already property waiting to be improved. For example, Twisted Gun Golf Course, in Mingo County, is a local icon of successful postmining development, cited in nearly every conversation I had with people who support MTR. Yet who is benefiting from this development? The golf course is managed by the three corporations who own the land and mined the area—and are still mining the area around the course: Mingo Logan Coal Company, Premium Energy, Inc., and Pocahontas Land Corporation, the largest landowner in West Virginia. And who plays golf there? When I visited the course, it was largely empty, with only a few cars in the parking lot; the place is relatively hard to reach from the most populated areas in the coalfields. Working people would have to invest an entire day off in just getting there and playing a game of golf, not to mention the expense of the equipment and fees. In any case, whatever wealth it does generate is going directly to the companies who mined the land and created it.

The biggest return on their investment is probably the public relations

Twisted Gun Golf Course, Mingo County, West Virginia, 2004. Photograph by the author.

value the place offers as an advertisement of post-MTR land use. As an exemplar of post-MTR development, Twisted Gun is an exercise in "spectacular accumulation" (Tsing 2005, 59). Its brilliant uselessness to the local community suggests that its actual purpose is economic showmanship; the golf course might create a new image of southern West Virginia in the eyes of potential outside investors in a way that more mundane projects fail to do. Twisted Gun spectacularly illustrates the power of property, a particular, objectifying relationship to land. I asked the clerk in the shop about the name of the golf course, and he told me a story that was evidently part of the lore about the place. According to this story, "during pioneer days," there were two (white) guys running from Indians. One of them was killed by the Indians, and the other took his friend's gun and twisted it between the branches of a tree. So, the clerk told me, the golf course got its name because "it's on that property." Referring to land as property, even in anachronistic contexts, reaffirms and eternalizes the white cultural relationship to land as private property. The reference to the land as property in combination with the frontier story of conflict with Indians evokes a racialized orientation to the American continent. It reflects the white cultural pattern of commodifying and mechanizing nature (Cronon 1991; Merchant 1980).

More practical examples of postmining development include, for instance, a veneer factory in Logan County, upscale housing developments, a small regional airport (perhaps servicing golfers headed to Twisted Gun), and the aquaculture experiments touted by Mike Whitt in Mingo County; several area schools and jails have also been built on reclaimed mountaintop mines.[4] But some of the "development" seems even more illusory than the golf course. When Jerry took me on a tour of some of these sites, he also showed me the "Earl Ray Tomblin Industrial Park," where the veneer factory is located. Most of the industrial park is still only potential, consisting only of a sign with the name and a paved two-lane road that winds through a reclaimed field and ends in a cul-de-sac. But the most serious criticism of this hegemonic perspective on postmining development is the one cited by Roger, that enough flat land has already been created by MTR for the entire city of Charleston to be relocated, the vast majority unused and undeveloped.

Even when the flattened land left by MTR is used for something, most postmining development does not directly benefit the majority of poor and working people in the area because it is perhaps coincidentally largely devoted to high-end projects like airports, golf courses, middle-class housing developments, and the production of exotic (to the area) commodities like arctic char and wine. These projects reflect the aspirations of local elites and coal industry boosters—like Twisted Gun, they offer a spectacle in which West Virginia leaves its backwardness behind. As I've noted, there is a tendency in discourse about Appalachia to equate the landscape of the region with the character of the people. Dark and dusty, the steep twisted mountainsides are forbidding, just as are the dangerous mountain men who live there. MTR, much touted for its capacity to correct the natural inefficiencies of the landscape, according to industry supporters can potentially straighten out the crooked landscape of the southern West Virginia coalfields. MTR converts wilderness to property, and in so doing it offers southern West Virginia a way to become more modern, more homogeneously American.

This belief in the development potential of postmining land indicates the dominance of white middle-class notions of nature, property values, and economic progress. The total erasure in pro-industry rhetoric of the forest and other life that is removed in a mountaintop mine exemplifies the objectification of nature. In a move strikingly similar to the ideology of European colonialism, in pro-industry discourse the land is often constructed as useless or empty. As Ned noted when discussing the Hatfield–McCoy Trails development initiative, "We had all this land in southern West Virginia, just sitting here." Likewise, Randy, a white DEP officer, described MTR as a process that has "taken areas that are un-useful and made them useful." He declared, "If I had a steep holler, I'd rather have useful land." In his view, mining was an unequivocal good. On postmining sites, he argued, "Once you plant trees, you have a better stand of timber than you had [before]." Because he is an environmental regulator, he said, if coal operators were not mining in environmentally responsible ways, he would not allow them to continue.[5] He also emphasized the distinction that many industry supporters make between "prelaw" surface mining and the regulated mining of today. It's true that surface mining was very destructive before the Surface Mining Control and Reclamation Act of 1977, he said, but today it is a "complex and scientific process." In these claims, the law provides a seemingly objective standard for truth masking the vested interests that were involved in shaping the law and its context (P. Williams 1991; Haney-López 1996; Harris 1993).

Pro-MTR discourse evacuates the land of meaning in a way that is reminiscent of the erasure of Native Americans in the colonizing discourse of early European settlers and of Manifest Destiny (Cronon 1983; Kuletz 1998). In this case, not only is much of the local forest-based *human* activity ignored but the nonhuman life on the unmined mountains is rendered completely invisible. Because most of the legal regulation of MTR is related to the Clean Water Act, which protects streams, valley fills are the subject of much of the controversy. This focuses attention on water quality and aquatic life instead of on the forest that is removed along with the mountain. Although this tactic has proven effective (in the short term) in legal battles for the environment, it narrows the argument in ways that can minimize the environmental harm of MTR.[6] As the mining engineer Chris told me, there is as much aquatic life in a "parking lot . . . [or] in [someone's] backyard" as there is in an ephemeral mountain stream. So, he implies, valley fills do not actually cause any environmental harm. Indeed, as his friend and coworker Rob added, "In fact there's a filtering process that could be a positive—is typically a positive factor. When you bring that water through all of that rock fill it's a cleansing [process]."

The county official Richard expressed his frustration with the environmentalists' objections to valley fills this way: "One thing . . . I've never been able to figure out is [why environmentalists care about] these small, small streams; there's no fish in those streams anyhow! You know, most of these small hollows you see, there's no fish in them. . . . There's not even any bait fish in them! I was born and raised here, and I know there's nothing in these streams."

The view of mining as creating useful land out of nothing was echoed by Horace, a white small-business owner in a coal-related field: "I don't think [environmentalists] realize what [surface miners] are doing. They're putting valley fills in the streams, sure, but they're not blocking off the streams. They are making flat land for accommodations later on, that we don't have in southern West Virginia. Flat land . . . If you go up on top of one of these [unmined] mountains, just walk up top of them, [you'll see there's] nothing there. [Coal companies] are keeping most of the material up there on the mountain. Piling it up, flattening the land, and when they do put these valley fills on the streams, they compact [the earth], they get them in good shape, they are terraced, [they grow] grass, deer . . . [the mining] didn't hurt a thing."

Horace expresses a modern, objectified relationship to land that echoes Locke's vision of European settlers improving the "empty" American continent (Locke 1690). Because I knew that he was a religious person, I asked Horace about the notion that it is morally wrong to destroy the mountains, as many environmentalists argue. He responded in terms that echo Manifest Destiny:

> I think God put that coal there, in the ground, as He did the timber
> . . . and there's nothing growing on that mountain . . . no one can
> build up in there; it's so steep that you can't access it, and the coal
> is in the ground. I believe it is there to get, and if they go about it
> in a good, clean workmanship manner, they should be able to get
> it. And if it's on certain people's property, and they would like to
> get it . . . out of there for income [they should be able to mine it].

Metaphors of a Racial Other

Just as industry supporters reiterate the discourse of colonization in justifying MTR, middle-class coalfield residents use metaphors of racialization to explain white poverty. Horace's disidentification from poor and working-class people affected negatively by coal mining is also represented in the remarks of the DEP agent Steve. Here, it is worth considering the race- and class-coded conversation that framed those remarks. In our discussion of the area's history, Steve's wife, Laura, called the area "Little Australia" as a way of describing the people who lived in the area "before coal." In other words, Appalachia was like a penal colony—inhabited by the dregs of the Empire. As the conversation continued, she described coal miners as "a special breed." Steve, who was not born into a coal mining family, as Laura was, frequently referred to a "coal mining mentality" and a "coalfield mentality," by which he meant that coal miners were prone to spending their money carelessly on the things they desired and getting into debt instead of saving their money or using it responsibly. Laura more sympathetically described the perceived local culture of poverty as "learned helplessness." The slippage in these comments reveals something about the complex articulations between race and class in American culture, where differences of culture and ancestry are metonyms for ostensibly economic distinctions.

Laura, whose father was a coal miner, told a story that illustrates how coal mining has particular symbolic signification. She described the diversity of the coal-mining community where she grew up and compared this to a later encounter across difference:

I went to school with the Japanese, with the Hungarians, with the Italians, with blacks, with different cultures, and we didn't know a lot of discrimination. I mean it was all just families there that were trying to make a living. A couple summers ago, [I took] a multicultural psychology class, and we were talking about race, and you know, all that, and our professor was a black woman [with a] very, very, [big] chip on her shoulder . . . But anyway . . . I said, "I haven't had your experiences, I don't know how you were discriminated against, and all of that," and I said, "but where I was born and raised, you know, everybody was different, it seemed like, everybody, and we just all tolerated each other, we all got along, we all played together." . . . There was a black preacher in the class, and he started laughing, and he said, "Yeah, Laura, down there where you're from . . . when everybody comes out of the mines, they're all black, ain't they?" (*Laughter.*) And I said, "Yeah!"

Here Laura describes a situation in which the "native whites," or white Appalachians, hold the place of the unmarked center of the diverse coal-field community, and she attributes her own experience of not knowing discrimination to the community as a whole. This assumption elides the perspective of the black families and families of other ethnicities whose experiences were potentially different from Laura's (Frankenberg 1993, 56). The statement claims an innocent white identity and naturalizes cultural and racial difference. As a part of a poor white family that didn't know privilege, she also couldn't know discrimination (Wiegman 1999). Strikingly, it is the black preacher who brings up the symbolic association of coal mining and blackness. Because of the specific material nature of coal mining, "they're all black." This could be interpreted as a support of her plea of racial innocence, or, reading against the grain, it is possible that the preacher's comment might have been a reaction to Laura's statement of innocent whiteness. Most interesting, however, is how Laura's comments move from a discussion of the diversity of her childhood community to a joke about mining and blackness.

Even when most coal miners are white, as they are today in southern West Virginia, coal mining is difficult to reconcile with whiteness because it so clearly reflects the abject body of labor. Phil Cohen argues that labor has two "phantasmagoric body images" that contribute to the production of laboring subjects (1997, 250). One of these is an idealized community of connection through labor; this is the site of masculine working-class pride in hard work that valorizes the whole community. The other is the embodied, abject condition of exploitation—the laboring body that is ultimately destroyed by the repetitive processes of work. The racialization of labor identifies the first, idealized, body with whiteness; white labor can and does read itself as the productive force behind the nation. At the same time, Cohen argues, black labor is identified entirely with the body. The labor process in this case is read as a disciplining of unruly carnal desires, instead of a creative or productive activity (250–51). White miners occupy an uneasy middle ground between these two images. While coal miners see themselves as essential contributors to the well-being of the nation and many proudly claim, "Coal is in my blood," at the same time, underground coal miners are lucky to make it to retirement with their bodies unbroken and able to breathe. Coal miners' pleasures—sodas, ATVs, video games—are portrayed as an undisciplined mismanagement of funds by critics of the "coal miner mentality."

The environmental effects of mining on communities further the abjection of the workers and their families. The feudal-like living conditions of the coal company towns, the almost unimaginable hardships of working underground in the "hand loading" era, the iconic image of the black-faced coal miner, and the epidemic of black lung all contribute to making coal mining particularly difficult to reconcile with white liberal citizenship. It is hard to think about independence when the house you live in is owned by the company you work for, hard to think about political liberty when the industry controls local and national politics, hard to feel modern and empowered when your body is bent and sickened by working underground. Even property ownership, the hallmark of American citizenship, is ambiguous when your lot is too small for a sewage system or is subject to damages caused by nearby mining.[7] In these regards, the historical conditions of coal mining serve as a clear example of the contradictions between democracy and capitalism.

The articulation of coal miners with the body and a lack of discipline is reflected in the analogies that some respondents drew between coal mining and concepts associated with blackness. For instance, in our conversation, Laura and Steve noted the trend toward "political correctness." Steve remarked, "There seems to be a phenomenon, you know, political correctness in this nation; you can't talk about lesbians, you can't talk about black people, you can't talk about ethnic groups . . . all that's politically incorrect. Got to be careful about what you say. But for some reason, it seems to be safe for states to talk about other states. . . . If you changed the label, if instead of 'West Virginia' it was, well, 'this black guy,' people would be upset."

Steve referred to the national stereotype of welfare-dependent Appalachians and noted, "There's people in every state that live on the dole. . . . West Virginians are normal people." Laura added, "That mentality exists everywhere, but here it's just a coal miner mentality. Everywhere else, it's the inner-city mentality."

Laura also noted that some of Appalachian specificity has its roots in historical facts. Describing historical coal mining conditions, she said, "I mean, the companies *owned* you." Discussing the famed Appalachian insularity, she argued, "When coal mining was established, people became suspicious of outsiders because they didn't know who they could trust, because the coal companies treated people so poorly. And if you talked about the coal officials, then you could lose your job. You could be thrown out and your family out of the company housing." For Laura and Steve, stereotypical

Appalachian behaviors result not only from poor Appalachians' welfare dependency, as expressed in the comparison of the coal miner with the inner-city resident, but also from a healthy distrust of the industry that so ruthlessly exploited its workers.

This high degree of racially coded exploitation isn't limited to the past, however, and it is not limited to people directly employed by the coal industry. As Paul, a small-business owner, noted, much of the conflict between Arch Coal and the community of Blair that ended in the closing of the Daltex mine in 1999 could have been avoided, but the company insisted on treating the people of Blair "like they were damn animals." Linda, an environmental activist, noting the prominence of Massey Energy Company in schools in their capacity as a "Partner in Education," said they were indoctrinating kids to become future "coal slaves."

These analogies illustrate that coal mining, both as a job and as a set of cultural conditions, is in some sense antithetical to ideal white American citizenship. Although unionized coal mining was part of the Fordist white-male-dominated "labor aristocracy," with high wages and good benefits, and even now is a popular career choice for young men coming out of high school, in other ways, coal mining occupies a symbolically low social position. Laura put it like this:

> As time passed, each generation seemed to raise their family and [think], you know, I don't want my kid to work in a mine. . . . Little by little, people have more opportunity, and it may have hurt the coal industry, because now you can't hardly get them to work in an underground mine. A lot of the boys say, well they ain't working in the mines!

And Meredith, a middle-class white woman and spouse of a mining-related business owner, described an insurance salesman and friend of her family this way: "He used to be a miner, but he's come up some."

Corporations as Ideal Individuals

While Meredith and I were discussing intraregional stereotypes about coal-mining communities, she brought up the racial prejudice of some members of her community. Her comments limn the symbolic American racial logic according to which middle-class whiteness is the standard to which black and working-class white people must measure up. Within this racial logic, "blackness" is the signifier of ultimate difference and is used as a metonym for white class differences, especially regarding coal miners. She described for me the racist attitudes of some of her acquaintances; "I [know some people] who just—if it's black—if [a] black [person] has it, it's going to be torn up pretty soon, and nothing's good about—you know what I mean?" Our shared whiteness forms the unspoken context for these comments, as signaled by the rhetorical "you know what I mean?" This question enacts our shared middle-class white status, mine as a researcher, hers as the wife of a businessman. In this white discursive space, businesses and corporations have special status as the ideal rational actors; in questions of moral worthiness they, like the ideal white citizen, are given the benefit of the doubt. This assumption of moral worthiness is not extended to workers. Most significantly, her comments offer an insight into the way that this racializing logic connects with corporate managerial practices. While she discussed her community's individual-level ideas about race and class, her comments also provide an example of the way that this logic of differential worthiness is reproduced on a different order by seemingly rational corporate practice. She described some of the difficulties coal companies face:

This is just a difference in people, Becky. You know as well as I do people are just . . . they're different now. Now it's very difficult if a job is coming [to an end] . . . because I've heard enough [coal] operators and enough people who are hiring these people to do the work. And they want to hire them and they try to keep them working. It's important. But if a job is coming to an end, [workers] will deliberately hurt themselves, so they can get benefits, or compensation. . . . So that makes it hard on the companies, and yet they do it deliberately! They'll cut off a finger! Can you imagine? But they do it! They do it. Now, this isn't—I guess I'm making it sound like it's an everyday occurrence. It isn't, but [the companies] have to be aware of it, because they know that there's always . . . enough of it has happened that they know it's a possibility. So they can't give these people fair warnings that the job is gradually slowing down—you have to keep it from them and that's awful. See, I look at that as a lie; I look at it from a different side. I want to know that my income's coming in; I look at it different. And to

think that men do that, just burns me up. . . . That one person can
hurt twenty people that were not guilty. OK, so I don't mean to
let you think that it's something that's going to happen every time
you close a job . . . but it happens enough that they have to be very
wary. And that's a shame.

Meredith names the injustice of companies' not telling their employees
when a mine is about to be mined out, yet she reiterates the corporate logic
that places the blame for this practice on workers who maim themselves
to gain unwarranted entitlements. She identifies with the innocent workers but perhaps cannot imagine the perspective of someone who would go
to such seemingly irrational lengths. For workers, the alleged actions of a
few—even the potential of this kind of action—justifies the companies'
habitual dishonesty with the hardworking, innocent majority of their
employees.

When it is a question of corporate guilt or innocence, however, a different moral standard comes into play. When I asked Meredith about Massey
Energy Company, she responded:

Anything that has to do with Massey—[activists are] always against
Massey. That's the biggest coal company around and they blame
them if a creek gets dirty, it makes no difference [which company
is at fault]—and now they can be guilty, too; there's times when
they're really guilty. But there's times that they get blamed for it
and fined for it when there wasn't a prayer that they were really
guilty. But people have to have someone to blame. Like suing
people? You know how that's become so popular? . . . It's the same
way with the coal companies, they have to have someone to blame.
And I'll grant you, sometimes they are [at fault] but not to the
point that poor old Massey's guilty of everything. And that every
coal operator is a crook. It's deceiving. It isn't true. Although some
of them are, just like some doctors are not good, OK? Is there any
difference in a coal company, the coal business, than anything
else? I don't think.

Several things come out of these comments. The burden of moral proof
has clearly shifted. Whereas a few bad workers can force all coal companies to lie to all their workers, a few, or even many corporate misdeeds, in

this formulation, are not enough to justify a uniform atmosphere of distrust for all coal companies. Workers must continually prove their worthiness; they are distrusted by default, but corporations are like ideal rational individuals; their mistakes are the exception. The phrase "poor old Massey" anthropomorphizes the company, underlining its ideal individuality. Significantly, Meredith compares coal companies to medical doctors who are saddled with expensive malpractice insurance due to too many lawsuits, but who are usually seen as benevolent. This emphasizes the good character of coal companies, who are just trying their best to keep "these people" working. Like doctors who are sued for bad outcomes even when they are the result of natural processes that could not have been prevented, Massey is victimized by people's seemingly irrational need for someone to blame. This comparison elevates Massey morally and naturalizes the damage caused by collapsed valley fills, slurry spills, and subsidence.

These comments represent a commonly espoused view of business-friendly culture, that union representation and benefits like workers' compensation allow undeserving working-class people to gain unfair advantages against corporations. Workers' compensation is almost never seen as merited in this context. Coal industry supporters typically describe a world where workers will go to any lengths to get out of working, where people concoct elaborate and painful schemes to defraud their employers. In this atmosphere, workers' compensation could only be considered fair if no one ever needed it; all accidents are presumed faked by the always distrusted workers. This identification of moral worthiness with ownership, and immorality with manual labor, bears traces of plantation logic. Like plantation owners, CEOs are represented as well-meaning and productive, but like the enslaved people on a plantation workers are "symbolically annihilated . . . either absent, trivialized, or condemned" for causing trouble (Eichstadt and Small 2002, 106). Unlike the shiftless and irresponsible workers, the coal company is constructed as a kind of super-citizen, an ideal liberal individual whose pursuit of its own interest ultimately benefits everyone.

My conversations with coalfield residents had nothing overtly to do with race or racism but were nonetheless shaped by the "capitalist, patriarchal and racist social order" in which we live (R. Johnson et al. 1982, 252). These conversations are a window into the inner workings of a cultural logic. For example, Meredith's comments express the logic that frames apparently rational corporate practice. Workers' compensation, in this cultural system, is articulated metonymically with affirmative action. Just as affirmative action is perceived as a threat to "fairness," or the historically accumulated advantages of white people, worker protection programs such as workers' compensation are a threat to fairness—in this case the right of corporations to accumulate capital. Like the historically amassed privileges of whiteness, the overwhelming dominance of corporate landownership in the coalfields is normalized through a discourse of private property, individualism, and free-market competition, which as Shawn pointed out constructs the geography of the region as problematic and the region's poverty as natural.

Massey Energy Company is a perfect illustration of the ideal corporate citizen. Massey's presence saturates the southern coalfields—everywhere you look you see Massey bumper stickers on cars and trucks, Massey T-shirts, and signs indicating that mines are subsidiaries of Massey. The major road through the coalfields is dotted with billboards advertising jobs with Massey. Union billboards sometimes also explicitly refer to Massey, as one that stated, "Massey's solution to problem solving? Throw money at it and hope it goes away. The UMWA's solution? Work together and solve it!"

Perhaps following the example of Walmart, Massey euphemistically refers to its workers as members, not employees. As of 2008, over 98 percent of the company's 6,743 "members" were "union free" (Massey Energy Company 2008). Massey CEO Don Blankenship explained in a 2005 interview with *WVInc.*, an online magazine, how labor laws and programs like workers' compensation hurt working people. Discussing the next political race he was interested in (he spent approximately $4 million to successfully defeat West Virginia Supreme Court Justice Warren McGraw in the 2004 election), he mentioned liberal Supreme Court Justice Larry Starcher and explained why he would oppose his reelection in 2008:[8] "He doesn't understand that by voting in every case for workers' comp claimants . . . or nearly every case . . . he doesn't understand that he actually damages workers. His love for the working community and for unionism actually hurts both and it hurts the state . . . and there's probably a few other individuals that far to the left that we need to defeat if West Virginia is truly to continue on a course of improvement" (WVInc. 2005).

Massey's "union-free" members face more difficult working conditions than miners in union mines. Massey miners are not allowed to take meal breaks during a shift and are frequently required to work overtime according to many of the people I interviewed. Several miners I interviewed reported dangerous violations of safety rules in Massey mines. The intensive labor practices make sense for the company because they serve the ultimate goal of increased profits. However, in public, Don Blankenship assumes the role of public benefactor. His campaign to defeat Chief Justice Warren McGraw in 2004 was managed through an organization called And For the Sake of the Kids. The group sponsored TV ads that accused McGraw of reducing the prison sentences of child molesters and other sex offenders. In his 2005 interview with *WVInc.*, Blankenship clarified his position on And For the Sake of the Kids:

> We didn't mean Tony Arbaugh [a person mentioned in the ads] and pedophiles. We meant that children are being forced to leave the state because they can't find a job when they graduate. They don't have the same educational level that the rest of the children in the country have. They don't have the same soccer fields or the same baseball fields available to them for sporting events. They just don't have the same opportunity that the rest of the country has, and they deserve that.[9]

Massey's good citizen posture is summed up by the slogan it displays on billboards around the coalfields, "Massey: Doing the Right Thing with Energy!" A 2005 television commercial for the company explained the corporate role this way: "Since the dawn of time, man has [sought to harness energy] to improve his life and the lives of others." In this ad copy, "man" clearly means the company, whose self-interested actions are implied to benefit all humankind. In this corporate discourse, the moral high ground is removed from those like Justices Starcher and McGraw, who claim to be

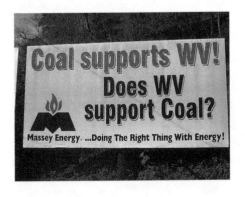

Corporate citizenship, Logan, West Virginia, 2004. Photograph by the author.

on the side of working people. As in *Atlas Shrugged*, the only true moral action is concerted pursuit of individual financial gain (Rand 1957). Don Blankenship claims this moral position at the end of his interview with *WVInc*. Asked what he would like people to know about himself, he responded, "I'm a good guy. I don't use drugs. I don't steal. I don't harass people. I live my own life in privacy. I do what I believe in, and anything I say I believe to be true" (WVInc. 2005).

The Micropractices of Whiteness in the Coalfields

Civilizational Standards in Housing

The family home is probably the most important piece of property for most Americans, aside from the extremely rich. Owning one's own home has been normalized as a common aspiration for all American citizens. Homeownership became the hallmark of the emergent middle class in the early twentieth century, when the ideal of a property-owning citizenry fueled the demand for inexpensive standardized home kits and provided copy for their marketing (Moskowitz 2004). The seemingly inherent value of the property relation was understood to create better citizens. "A family that owned their own home became part of the community in a way that renters did not, as their own 'welfare and progress' became inextricably linked to that of the community" (140). According to this ideology, owning property established an official and singularly meaningful connection between owner and place. Simply living in a place is not enough of an investment; property owners are the inhabitants who matter. As President Herbert Hoover put it, "It is mainly through the hope of enjoying the ownership of a home that the latent energy of any citizenry is called forth" (quoted in Moskowitz, 140). With the advent of inexpensive mail-order homes, new members of the middle class were able to become formally equal to the wealthy.

If proper American citizenship is related to homeownership, coalfield property relations once again fall short of that ideal. The connections between the economic dominance of the coal industry and the substandard living conditions of some local residents are disguised through the language of private property. Recall what Steve described for me as the "legal snafu" that has prevented some coalfield homeowners from having proper sewage disposal, because former company houses were built on lots too small to accommodate a septic tank. Ironically, by granting individuals the right to own property, the sale of these company houses dismantled the community power that existed during the coal camp days. In those days, companies provided services for the workers and their families. As nominally independent homeowners, the inhabitants of the former coal camps have too little leverage to control the environment outside their small holdings. Perhaps those who rented from the company were better off, since the company in those situations was still involved and maintained a minimal interest in the upkeep of the property. However, those who rent are subject to eviction.

Steve described the community of Rum Creek, a former coal camp, recently dissolved by its owner, Dingess-Rum Land Corporation, which had evicted all its tenants and offered them spaces to rent in a nearby trailer park. The company was now leasing the land where the community of Rum Creek had been located to a coal company for mining. He said:

> We both knew people who'd rented a house [from Dingess-Rum] for a long time, you know, and they'd rented it for fifty dollars a month. And they'd been there all that time. But they improved the property, they remodeled, they put money into it, they developed— they built them a nice home.

Laura added, "*If* they were working people." Laura and Steve were former residents of Rum Creek; they had lived in a trailer there. As middle-class professionals, they were able to buy a house when they were forced to move, unlike many of the other residents of Rum Creek, many of whom, if they could afford a trailer, moved to the trailer park near the mine site.

Jerry, who took me on the previously described tour of postmining development sites in Logan County, gave me a different middle-class perspective on the evictions. In his view, the residents of Rum Creek were better off now that they were in trailers.[10] To my questions about the eviction, he answered no, the people didn't have any choice about moving because they rented from the company that owns all the land. But since they were moved into trailers, they were now in better houses than they had been before. He described the former coal camp community this way, "You can be poor, but you don't have to be nasty."

Possibly the most common type of residence in the coalfields is a trailer. At around twenty thousand dollars for some used models, and with lots of fancier designs available, a trailer offers a relatively easy way to buy a house. However, many of the trailers are old, and in fact most of the available rental units in the area are old trailers. A kind of disposable housing, trailers are not constructed to stand up to flooding or to last a family for generations. Although a new double-wide can be a quite comfortable place to live, many old trailers in the coalfields are in very poor condition.

In addition, trailers are loaded with shameful cultural symbolism. The private property of trailer dwellers doesn't immediately warrant respect. For example, a community in Mingo County successfully sued Massey Energy Company in 2004 for property damage due to subsidence and for providing bad water to replace lost well water. I attended the penalty phase of the trial, where the defendant's lawyer minimized the levels of damage claimed by one of the plaintiffs because her house was a double-wide trailer that was not in good condition to begin with. The lawyer implied that given the poor original condition of the property, the fact that Massey's undermining made it uninhabitable was less significant. While cross-examining this plaintiff, Massey's defense lawyer focused his questioning on the woman's front porch. The plaintiff described how undermining destroyed the porch, forcing the family to use a plank to access their front door. The lawyer's point seemed to be that the porch was unstable to begin with, because her husband had built it himself instead of hiring a contractor.

The example of the Massey trial also illustrates the fragility of individual rights to property and security in the coalfields. The plaintiffs were property owners, but the coal mine under their houses ruined their wells and in some cases damaged their homes. There was no city water available, and until it could be brought to the community Massey provided water in large plastic vats to replace the water from the destroyed wells. Several children were made very sick, possibly by E. coli bacteria in the replacement water. Many of the residents simply mourned the loss of the free, fresh springwater they were used to drinking from their wells; they did not want city water. In the lawsuit, the company's defense lawyers tried repeatedly to establish that the community was better off now that they had access to city water. So much better off that the fact that this switch was made against their will was negligible.

Whiteness, Housing, and Hygiene

The problematic character of the people in the former coal camps is symbolized by those paradigmatic sewer pipes leading straight into the river. Both local residents and outside commentators frequently invoke the failure of coalfield dwellings to meet modern standards of hygiene as a testament to the residents' difference. The connection between indoor plumbing, hygiene, and modernity has been consistent at least since the early twentieth century, when the modern bathroom was promoted by manufacturers and social workers as a new necessity. Along with the "standard" fixtures of toilet, bathtub, and sink, a new standard of hygiene and privacy was established. Moskowitz describes the new standard this way: "While bathing could be achieved anywhere with a basin of water, having a designated space within the house for the purpose of washing oneself lent an importance to the act that it might not have otherwise had and allowed for a greater degree of privacy for actions increasingly constructed as personal" (Moskowitz 2004, 67).

Indeed, the new cadre of social workers who began managing the urban poor and European immigrants at the turn of the century saw the modern bathroom as essential to a hygienic American lifestyle. As social worker Florence Nesbitt judged in 1908:

> There should be toilet facilities in good condition with a door which can be locked, for the use of the family alone; running water in at least one room in the house besides the toilet. A bathroom is highly desirable and should be included wherever possible. It may be considered essential in families where there are a number of older children in rooms which would not otherwise permit the necessary privacy for bathing. (quoted in Moskowitz 2004, 67)

Marketers and social reformers constructed bathrooms, formerly a luxury for the rich, as a new necessity for the middle class and at the same time promoted new standards of both personal hygiene and housekeeping. As a 1903 text on physiology and hygiene instructed, "The one important condition of public health is cleanliness . . . [including] clean private residences . . . and cleanliness of person and clothing" (quoted in Moskowitz, 66). According to Moskowitz, "Lessons about cleanliness were directed at all socioeconomic groups, but high standards of cleanliness were often seen as the marker of middle-class values, and were aggressively introduced as aspirations for the working classes" (66).

These new standards of public health and hygiene have greatly contributed to the increased life expectancy of most Americans throughout the twentieth century. However, these new objects and practices were overdetermined; they were used and understood within a cultural context heavy with racialized understandings of cleanliness and dirtiness. Race and class distinctions were entrenched through differential access to cleanliness; segregated neighborhoods and tenement housing limited the ability of working class people, immigrants, and African Americans to achieve these new standards (Tomes 1998).

A British soap advertisement from the late nineteenth century exemplifies the racially coded nature of these practices and aspirations. In the first of the advertisement's two frames, a white child washes a black child in a tub. In the second frame, the black child has emerged from the bath to find himself "whitened" up to his neck (McClintock 1995, 213). This association was repeated in early twentieth-century Texas, where Mexicans found themselves in the "ethnoracial middle ground between Anglo Americans and African Americans." For instance, Mexican children "could attend Anglo schools if they were 'clean,' which was often a euphemism for 'white.'" Researcher E. E. Davis said in 1925, "The American children and the clean, high-minded Mexican children do not like to go to school with the dirty, 'greaser' type of Mexican child" (quoted in Foley 1997, 41). The connection with cleanliness and whiteness was explicit, as a Texan of mixed Anglo and Mexican parentage made clear by noting that "'clean Mexicans with Spanish blood and fair complexions' were allowed to sit next to Americans" (quoted in Foley 1997, 42). These articulations between cleanliness, moral worthiness, and race persist today, as evidenced by one of my respondents, who commented, "Some of my best friends are black. And they are just as nice and just as clean and just as smart."

These associations became naturalized historically through overlapping, redundant repetition of similar associations and oppositions. Rapid shifts in upper-class American dinnerware led from trenchers and natural-colored ceramics in the Revolutionary period to white porcelain by the 1830s. This whitening of material culture, including houses and gravestones, as well as dinner plates, parallels the increasing rigidity of American racial categorization at the same time (Heneghan 2003, 8–11; D. Wilson and Beaver 1999). The symbolic association of middle-class morality, whiteness, and cleanliness was again reinforced in the marketing of the new, cheaper bathroom fixtures in the early twentieth century, where "The glossy finish of enamel, especially in pure white, connoted not only hygiene but also well-designed domestic goods, like the porcelain with which it was associated" (Moskowitz 2004, 70). Moskowitz also notes that "until the mid-1920s . . .

most . . . companies produced wares in white only. In fact, competition between companies arose over the issue of who could create the whitest enameling powder" (71).

Paul, a former miner and small-business owner, was raised in a postwar coal camp where his father worked as a miner, and he describes coal camp life from the perspective of a mining family. His stories demonstrate a preoccupation with cleanliness that reflects the way modern standards of hygiene were "aggressively introduced as aspirations for the working classes" (Moscowitz 2004, 66). Paul grew up in a family of eight that lived in a four-room coal camp house. He remarked on the degree of control the company had over the daily life of the workers, right down to aesthetic choices: "[If] you worked for Boone County Coal, well, then, [you lived in a green and white house because] they painted all their houses green and white." As a boy, he shared a bed with his two brothers. They had an outhouse, and while the girls of the family were allowed to use what they called the "bean pot" at night, the boys had to go outside. One of the boys' chores was to carry the girls' bean pot out to the toilet in the morning. During the winter, it was often so cold in the house that the pot would be frozen solid in the morning. On nights like these, the family would have to sleep with "twenty pounds of quilts and blankets on top of [them]." Referring to the crowded conditions in the house, Paul said, "But that would all be unsanitary, today, to try to do that. Three boys in the same bed, and then in the living room, that rollaway bed kicked out and two girls in there [and] on the other side of the room there's a curtain you pulled."

As a boy, Paul helped take care of the house and the small farm that helped feed his family:

> My dad worked evening shift . . . [and] us kids, when we got up in
> the morning, . . . we would have to . . . poke a fire in the grate, and
> then we'd take out the ashes, and we'd bring in the coal and the
> kindling, and everything . . . then in the kitchen we'd have to build
> up a fire in the coal stove, 'cause Mother cooked on a coal stove.
> And then we'd go up on the hill and we'd gather the eggs, slop
> the hogs, feed the chickens, feed the horse, and curry the horse;
> I mean it had to be done. I mean that religiously. And then when
> the hog would jump up on the slats, they'd throw pig shit on you,
> so you'd have that on you . . . and then you've got to go to school
> with all this. Then when you got on the bus . . . the kids would

poke fun at you, "Pew, I smell pig shit!" And you'd feel about that
tall. And us three brothers would go back there, and just hang our
heads, and just sit down there, and we'd *take* that abuse. Well,
I always said, with my boys, that they'd never have to do that.
When they got ready to get on the damn bus, they could smell
like a human being.

He also described the ritual of his dad's bath after work:

My dad would come in from the coal mines on evening shift, and
we had a round . . . number three tub. Well, [we] had it figured
to a science, now, how many buckets of water to put in it, and it
would be there all evening, all night long, a-boiling. . . . We had a
piece of plywood . . . that we laid down [on the floor], and we had
a washtub . . . just like the one we had . . . on the stove. . . . we
called that one the blacken tub and this one here was the clean
tub. Then this tub sat down here on the floor . . . [with] six or
eight inches of water in it. He'd leave the house at three, and then
they'd go in at four and get off at twelve [so] by the time he drove
to the house it's like one-fifteen or one-thirty [in the morning].
He'd come in and my mother'd fix him a bite to eat. Whatever *we*
ate, she'd heat it up for *him*. There wasn't no damn microwave. It
was on that coal stove. . . . [If we had to get up in the middle of the
night to go to the outhouse, we'd see and wonder,] why the hell's
he bath [*sic*] like that? He would get down on his knees; he'd wash
his head, wash his face, wash his neck, wash underneath his armpits,
wash his arms, so he'd have the upper part of his body washed.
Then he'd take his pants off, and then he'd get into the tub. And
then he'd wash the rest of himself.

This detailed story of his dad's bath was part of a tale he'd evidently told
many times, which ended with Paul working up the nerve to ask his dad
why he bathed that way, and his dad answering, "Hey, boy, you never wash
your damn face where you wash your ass!"

Paul's memories of growing up in a coal camp house indicate the importance of hygiene and cleanliness to miners and their families. The temporal
narrative of development inherent in these standards is indicated in Paul's
comment that the conditions of his boyhood home would be unsanitary

today. For white coal mining families, the dirtiness of the job, and the near impossibility of maintaining modern standards of hygiene in coal camp houses, perhaps uncomfortably evokes the experience of African Americans and Mexican Americans who have been racialized in terms of cleanliness, hygiene, and housing.

Meredith is married to a white businessman who once worked for a coal company. In her description of her short period of residence in a coal camp where her husband worked, Meredith remembered the many inconveniences she experienced there. Despite her liberal political orientation, her comments are structured by deeply entrenched cultural hierarchies of race, gender, and class. Because her story is expressed within the prevalent discourse of differential development and worthiness, it reaffirms these hierarchies at the same time it questions them. Rather than determining her "true" attitudes as opposed to her expressed values, what is interesting is how underlying cultural notions of race, class, and cleanliness are articulated in the experiences she recounts.

Meredith explained that when she and her husband were newly married, he didn't want her to live in the coal camp where he worked, so he commuted weekly and she stayed with her parents. She said, "He thought I couldn't live in a coal camp, that I just couldn't do that." But she grew frustrated with only seeing him on weekends and eventually insisted that they move to the coal camp together. She described the house they lived in:

> [It was a] four-room house, I mean *four rooms*. A door in each corner, so the kids could ride their tricycles around ... an outdoor toilet. I was lucky, [my husband] put water into the house for me. So I could have a washing machine, and I could have water running and I didn't have to carry it. Now that's ... primitive! These houses were very—they put them up in a hurry. Because this job started, and they built these houses, and I'm sure the lumber was green and pretty soon you have cracks in it; it was ... kind of rough.

One thing these comments do is to illustrate how an economically rational corporate practice—the quick and slipshod construction of houses—can begin to shape perceptions of the people living there. The coal camp houses were primitive and rough, but Meredith stressed that most of the people that she met in the coal camp were good people. "You made good friends; you helped each other out. I met some of the nicest people I've ever met there. And some that I—I don't remember meeting anyone that was really bad, but some that I wouldn't have wanted to have been around a lot. But everybody was helpful; I didn't have any complaints." Meredith may have felt the need to emphasize that most of the coal camp residents were nice, because according to the logic of differential worthiness the niceness of the poor is always already in question. Later, when we were discussing intraregional stereotypes of coal miners, the articulation of color and class logics in structuring the cultural signification of coal mining found expression:

> Daddy told me when I was moving to the coal camp, "Meredith, you have to be careful, they are just different." Well, they *are* different, but that doesn't mean that they are black—uh—bad. It's kind of like black people and white people. Just because they're black, they're not all bad. My perception of a lot of things is probably different from a lot of people's . . . Like the example I just used with black. See, I have no trouble, I have no problems with it, but [I know people] who are just really terrible when it comes to black people, and the same way with coal miners or whatever.

Manufacturing Whiteness: The Funeral Parlor Aesthetic

The style of house construction and decoration that I came to call the funeral parlor aesthetic exemplifies the middle-class ideals of modern citizenship, cleanliness, and privacy. The poorest coalfield communities, the ones most likely to suffer from chronic unemployment and drug abuse, are former coal camps where residents often live in the remains of company houses, or sometimes in trailers. These old coal camps, and other working-class communities like the one that sued Massey, with their ramshackle porches, old double-wides, and stubborn attachment to well water, are exactly what the middle-class funeral parlor aesthetic is trying to leave behind.

Most of the old coal camp houses are gone now. But some still exist— one of which I considered renting while looking for a place to live during my fieldwork. Like many former coal camps, this neighborhood consisted of a one-lane road that followed the curve of the mountain hollow to its end near the crest of the hill. This particular place was a nearly intact coal camp, crowded with houses tightly placed on both sides of the road. These were houses the company had sold to individuals. The houses themselves ranged from literally burned-out husks to recently painted neat dwellings with flowerbeds planted all around. It was a mixed neighborhood, in terms of race. In terms of class it ranged from destitute to getting by. The very poor lived in barely standing structures next to houses that had been "kept up" by people with some access to resources.

The one I was to rent was covered in redbrick-look "tar paper" siding. The house had three rooms (kitchen, living room, and bedroom, plus a tiny bathroom), a big front porch, and a yard. When I tried to open the bathroom window, the glass almost fell out of its frame, and in the bedroom the window was nailed shut. This house was prominently located on the curve of the road and thus had a larger yard than the neighboring houses. The yard was scattered with debris, as was the interior of the house, including bits of trash, glass, cigarette butts, dead insects, nails, and other objects. There was a bucket of murky water sitting outside the back door, breeding mosquitoes. The kitchen was clean, but the carpeting in the living room and the bedroom was filthy. Additionally, the landlady told me, most days the yard would be dominated by the sounds of the limestone quarry, located just across the ravine. The rent was $250 a month, and the previous tenants had either left suddenly, unable to pay the rent, or they had been evicted—the landlady was not precise. The landlady, who was white, lived in a house up the road. Along the road, there were signs handwritten with markers on cardboard warning that the neighborhood was under video surveillance. When I asked her about these signs she told me that they were meant to deter drug dealing and that there were really no cameras.

After a period of initial excitement about having found a house in the country, I finally decided I couldn't possibly live there with my two-year-old daughter. I came rudely and abruptly face-to-face with my class position and my own nonnegotiable middle-class standards regarding my child. But I had plenty of other reasons for my decision. Examining the conditions of very poor neighborhoods was not part of my project. The middle-class people whom I wished to interview might be put off by such an address. The place was not easily accessible, it would be difficult to live there alone with a two-year-old . . . and I was pregnant. Lots of reasons. I eventually realized that the coal camp house was not the place I had imagined renting. From my apartment in Santa Cruz, I had imagined living in a rural house in West Virginia, like the ones I knew in Greenbrier County while growing up, surrounded by peaceful nature and with lots of privacy, where my two-year-old could run around naked and play in a wading pool. In the coalfields, I found that such a place did not exist, at least not to rent, not on my budget. So my search turned away from my dream of rurality to places that were simply clean, solid, and safe enough for us to live in. This is when I discovered the funeral parlor aesthetic.

Eventually someone told me about an apartment located in a trailer park that was owned by a man who reportedly "didn't put up with any nonsense." Although it was not technically a trailer, the apartment was built along the lines of one, with the rooms all in a row. It had only three windows, all on the side facing the parking lot, although the other side faced a meadow. It had a kitchen/dining/living room area, a bedroom and a bathroom. It was clean, solid, and modern; a family friend pronounced it "delightful," and we lived there for a while. It had white vinyl siding and the stairs outside still smelled of fresh-cut lumber. Inside, the apartment was decorated with framed art, the frames screwed permanently into the wall. The one-sided-window feature meant that there was no air circulation at all, and since it was sweltering outside, there was no choice but to use the central air-conditioning. The apartment seemed to be completely cut off from the surrounding land. There were beautiful views of meadows and mountains all around, but all that was visible from the apartment was the landlord's garage and the parking lot.

Inside the bedroom closet, the landlord had placed a laminated list of twenty-four rules, including:

Garbage is to be placed in plastic bags and placed in cans.
No repairing or overhauling of cars will be permitted on property.
Outside storage of miscellaneous equipment of any kind is
    forbidden.
Each rental unit is to be kept neat and clean with no storage of
    bottles, boxes, cans, tires, etc. outside rental units or on porches.
No loud parties will be allowed at any time nor will loud radios or
    other excess noise be tolerated.
Positively no alcoholic beverages are to be consumed outside
    rental units. These must be confined to inside the unit.
No outside clotheslines.

This exhaustive list of rules constraining the behavior of renters at the trailer park illustrates some of the rights a person gains as a property owner.[11]

It interpellates renters as people in need of correction, who must be warned against their potential "trashy" behavior. In the context of the coalfields, it is in the interest of the landlord to distance his rental units as far as possible from the house in the coal camp, even if that means sealing them off from the environment and attempting to strictly control all outward signs of the renters' existence. The oddly designed apartment can only make sense in the shadow of the tar-paper coal camp shack.

The funeral parlor aesthetic is widespread in the coalfields (as it is elsewhere in the United States). Often there are only small indications of it—like the fact that in Madison the florist sells mostly artificial plants and flowers, no fresh flowers except a few corsages for special events. One of my interviews took place in a middle-class neighborhood in Madison, across a table decorated with a beautiful lavender flowering plant. I asked what it was, and my respondent told me he didn't know and that it was fake anyway.

This aesthetic is most obvious in the newer houses and housing developments that dot the coalfield landscape. One type of new construction that exemplifies the look is a large new house of the design called "New American," built in the center of a wide, empty lawn (ePlans.com 2006). This house would have a sidewalk from the driveway to the front door, and the door would very likely have yellow-orange textured glass on either side, in the style of a well-funded Protestant church. Or this house might be found in a new upscale development, like one in Holden, which is being constructed on a reclaimed strip mine.

One of my respondents lived in such a house. His home office was off to one side of the entryway. The house was immaculate and cooled by central air. Tiled stairs led to a carpeted hallway, and I could see a perfectly clean kitchen through the hall. Large green artificial plants decorated his office. It was clear that this couple worked hard on making their house look as though no one lived there; everything was in place, spotless, and chilled. My respondent had worked his way into the middle class. He was the son of a coal miner; his brother worked in a mine and lived in a trailer not far from my apartment.

Houses like these are so far from the coal camp as to be in a different space entirely. They seem to have very little interrelationship with the surrounding environment—isolated in a large lawn, with minimal landscaping, they need air-conditioning to remain cool in summer. Or built on a reclaimed strip mine, they occupy a manufactured spit of flat land amid mountains all around. This widespread trend toward sealed, cooled,

Housing development on reclaimed strip mine, Logan County, West Virginia, 2004.
Photograph by the author.

immaculate, and isolated houses begs for interpretation. Among other
things, it seems to indicate a problematic relationship with nature, which
is understandable in the coalfields, although of course this trend is not isolated to the coalfields. There is a sense of unease among those people who
live in the hollows and the mountains, perhaps due to the instability of
nature in the shadow of coal. In less populated places in the coalfields, too,
there is often unexpected danger—auger holes left from abandoned longwall mines, undetonated explosives from old strip jobs, buried trash that
someone didn't take to the dump. Often the most beautiful places in the
coalfields bear scars of flooding, mining, and other trouble. Rum Creek,
outside Logan, is one of the prettiest creeks I've ever seen, but when I was
there debris from the recent flood hung in the trees along the creek, resem-
bling wet toilet paper and giving the creek the aspect of a sewer.

A 2005 television commercial for Massey Energy Company showed a
housing development in a place unrecognizable as West Virginia. With the

voice-over describing how man strives to harness energy to improve the world, the camera panned across a community on a manufactured lake. Each identical street in the development extends into an identical strip of water, in a perfect scenic manufactured community. This ideal neighborhood is a literal utopia of middle-class comfort, and coal makes it possible. Meredith's comment about her family friend could be an expression of the dreams of the regional elites pushing MTR as a source of postcoal development. It used to be a coalfield, but it's come up some.

But another trend in coalfield house construction frustrated many of the middle-class people I met. Frequently new, expensive houses are built in very unlikely places, such as immediately next to an old tar-paper-sided coal camp house or beside a flooded-out trailer park. For some people, evidently, neighborhood property values are less important than attachment to place, which is frustrating to many who wish there were more zoning laws to prevent this kind of mixing and find it an indication of a "coal-mining mentality"—that people don't think ahead to their house's resale value. The incongruity of elaborate new houses (I saw one with a tennis court) located immediately next to poverty-level dwellings indicates that there is more going on in the coalfields than the funeral parlor aesthetic. Local specificity in the form of attachment to place and other idiosyncrasies continues to grate against the desire for placelessness.

How we imagine the land has real consequences for both human and nonhuman life. These imaginations become "naturalized" through their echoing articulation with other cultural forms and by their materialization in the built environment. Abstract space, a kind of placeless commodity deployed by capital, becomes visible in the enactment of MTR on the land. Specificity, embodiment, and labor become signifiers of suffering in this powerful discourse of modernity. In this way, these signs of life are readied for sacrifice in the name of development. Our relation to the land is expressed in a myriad of small everyday ways that structure our experience of nature and of our own "corporeal vulnerability" (Thrift 2008, 242). The places we inhabit shape these experiences and are saturated with meaning by virtue of their embeddedness in intersecting and mutually productive cultural formations.

Extremely conscious of their place in the national imagination, many coalfield residents strive to represent themselves as worthy national citizens. The terrain of this struggle is often the particulars of relating to nature through everyday life. While environmentalists and community activists argue that coal is destroying Appalachia, others seem to desire an individualistic, placeless, modern American standard of living. The overwhelming dominance of the coal industry is naturalized through a racially coded discourse of individualism, private property, and free-market competition. In this discourse, corporations are ideal liberal individuals and citizens. While these ideal corporate citizens rapidly transform the Appalachian Mountains into "improved" flat land, some coalfield residents aspire to the same kind of citizenship.

Coal Facts

*C*OAL *FACTS* IS A BIANNUAL SUMMARY of coal production information released by the West Virginia Coal Association. The 2006 edition included two remarkable yet at the same time quite ordinary advertisements. The first was for Walker Cat, a heavy machinery equipment company, and featured a photograph of a white heterosexual couple dancing. The woman is wearing a tight, strapless red evening gown; the man is dressed in the uniform of an underground miner, reflective stripes marking his arms, legs, and groin. The ad copy reads as follows: "West Virginia has the hottest coal on earth, and the world's greatest coal miners. And now with today's modern coal industry, West Virginia stands poised to be a global energy leader. It's time for West Virginians to take a stand and answer America's call for energy and security. Shall we dance, West Virginia?"

The second ad consists of a mostly white page with the company name of the Pocahontas Land Corporation spelled out in an arc above the company logo, a stereotypical representation of the head of a Native American woman in profile, wearing a feathered headband and braids. Underneath the text continues, "a subsidiary of Norfolk Southern" and gives the company mission: "Our vision is to be the most responsible, innovative and successful owner and manager of natural resource properties."

The first ad shows a middle-aged couple in an unnatural, if legible, pose, and the second features an overworked and seemingly retrograde cliché. The painful awkwardness of both seems to stem from their too-naked appeals to sentimental and traditional articulations between sexuality, nation, and land (Probyn 1999). The white woman is West Virginia, on the verge of a beautiful experience with the miner. This is not a rape, it is a seduction. The ad could be an enactment of the popular bumper sticker proclaiming "I Love Coal." Calling on West Virginia to embrace the domination of coal, to read this relation through a lens of "natural" social forms, the industry tries to drown out the voices of activists who interpret this relationship differently. The Pocahontas Land Company ad features not a white woman

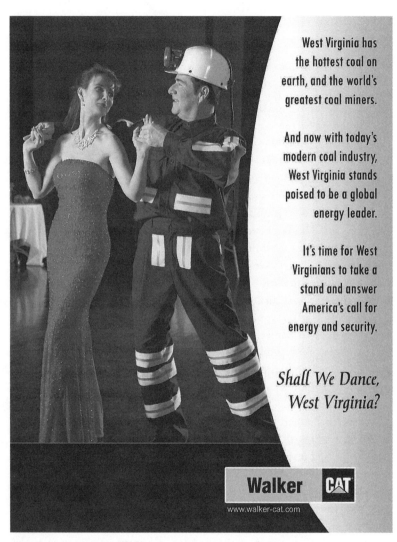

West Virginia has
the hottest coal on
earth, and the world's
greatest coal miners.

And now with today's
modern coal industry,
West Virginia stands
poised to be a global
energy leader.

It's time for West
Virginians to take a
stand and answer
America's call for
energy and security.

*Shall We Dance,
West Virginia?*

Walker CAT
www.walker-cat.com

*Walker Cat: the exciting modern heteroromance of coal (West Virginia Coal
Association 2006).*

Pocahontas Land Corporation
(West Virginia Coal Association 2006).

a subsidiary of

OUR VISION IS TO BE THE MOST RESPONSIBLE, INNOVATIVE
AND SUCCESSFUL OWNER AND MANAGER OF
NATURAL RESOURCE PROPERTIES.

but a woman of color. In this case, the image speaks less about a relationship, however structured by power differentials and gender norms, than
about a romanticized object, the land as fertile natural resource. These
images signal the ways in which "the economies and political relations of
nations are libidinally configured" (Tadiar quoted in A. Smith 2005, 8). If
in these ads the Native woman represents the land, the miner "*literally* bru-
talizes her, while *symbolically* brutalizing the white woman" who represents
the abstract social body of the state (A. Smith 2005, 22).

These gendered and racialized representations of West Virginia (as land
or as a social body) animate various relationships between the state, landowners, and the coal industry, workers, and the national community. As
the Walker Cat ad specifies, this network of relations affiliates with modernity and American national identity. The national symbolic is on view as
the excessively heterosexual couple (sexual difference exaggerated through
divergent styles of dress) prepares to enact the romance of mining. At the
same time, the text locates West Virginia at a pivotal and vital location
between the nation and the globe; the state stands ready to be a global
energy leader, at the same time as it answers America's call for energy security. The relationship here offers freedom from dependency on "foreign"
oil. This is accomplished through a distinctly modern industry; the yellow
reflective stripes on the spotless mining uniform signal that the dancing
miner has superseded the dirty-faced miner in overalls, covered in coal dust.

The dancing miner seems ready to lead his partner into a beautiful future
of prosperity and security, but the Pocahontas Land Company ad signals
the history that is elided by the "absent referent" (Adams 2000, 55). Native
American women have long been used to symbolize both the American

landscape and the wealth of natural resources. Whether the object of a seduction by plow or a rape via dragline, the land has been identified as a female with brown skin in American culture: "The great brown earth turned a huge flank [to the sky] exhaling the moisture of the early dew.... Deep down there in the recesses of the soil, the great heart throbbed once more" (Norris quoted in Merchant 1995, 146). Constantly invoked in place-names, manners of speaking, and cultural idioms, this absent referent signals how the violent history of gendered and racialized domination of people and territory haunts American national identity. The Pocahontas Land Corporation's name and logo exemplify the use of Native American culture and images to represent American history (Kazanjian 2003, 12). Represented in this stereotypical form, the head of Pocahontas represents the "vanished" Indian who seems to have paved the way for modern America and whose evocation consolidates a modern American identity (Coombe 2001, 32). Indian place-names, team logos, and claims to "Indian blood" are seemingly the "trophies" of colonialism (Garroutte 2003, 55).

The collective project of modernity is expressed through civilizational discourses, embodied dispositions, and the condition of the physical environment. These gendered dispositions set up a "*whatever* point" to interactions with nature that tell a conventional, romantic story of what mining means and lend charisma to the technological practices of underground and surface mines (Berlant 2008, 266). The connections between these discourses, embodiments, and environments are elided through the disassociating discipline of economics, politics, and culture. At the same time, these buried connections endue political-economic logics with affective power. MTR, as an example of the restructuring of nature for humanity, is made meaningful by its placement within the discourses, dispositions, and material practices of modernity.

Coal industry supporters reiterate the colonial imagination of the land as empty, useless, and waiting to be improved. This is the abstract space that Lefebvre sees as intrinsic to capitalism, the land objectified and accounted for as individual private property (LaDuke 2005; Lefebvre 1991; James Scott 1998). As the object of market exchange, land can only be improved by development (Blomley 2002). Peter, a white middle-aged educator, reflects this view. He and his wife live in a middle-class neighborhood in Madison. I asked him if he agreed with one of his colleagues, an outspoken critic of the coal industry and MTR. He answered, "Oh, I'm much more mellow. We fight. I say, take the mountains off." When I asked him to explain, he said, "Why do I say that? Well, I guess I'm selfish. There's jobs created by [MTR], that makes economic prosperity here and that makes property here worth more money, this house worth more money. . . ." This perspective reflects an articulation between individual self-interest and private property; i.e., the only way he will be affected by MTR is by its effect on his property value, which he defines in relation to the availability of market opportunities. He says this is selfish because he does not share his colleague's selfless (and misguided) devotion to the environment. His comments reveal a cognitive separation between his own self-interest, as reflected in his personal property, and the nonhuman nature threatened by MTR, which he considers a mostly altruistic concern of environmentalists like his colleague.

These comments express the traditional white middle-class orientation to the environment, which understands nature to exist only in the wilderness, apart from everyday modern life, where market-driven technological development always improves and adds value. Additionally, industry and economic development are naturalized as inevitable—those areas that are undeveloped are backward, left behind. For Peter, the town of Madison benefits more from the progress represented by economic activity, whether that be in the form of fast-food franchises or mining, than by the preservation of useless forest.

People who speak for the industry seem to be incapable of imagining an environmentalism that goes beyond the traditional Western environmental ethos of wilderness conservation. This lack of comprehension is reflected in their comments indicating that environmentalism is impractical, altruistic, and romantic. Many coalfield activists identify as "concerned citizens," not environmentalists, for this reason. "Tree hugger" is an epithet repeatedly applied to anti-MTR activists, but it is an inadequate description of their cause. The term suggests a misplaced affection for an alien life-form—a kind of betrayal of humanity (and is also syntactically close to a racial epithet also traditionally used with negative meaning). As Max, the white owner of a coal-related business, put it, environmentalists want to "save crickets and let people starve." Ella, the wife of an underground miner, said environmentalists would "kill a man to save a tree." She recounted how she expressed her anger in online chats on coal-related issues, remarking, "The people that are posting against it are from California and Illinois. Do you want me to pick a tree over my child? . . . Do you love a tree that much?" These comments reflect the naturalization of the place's identity as a coalfield; in other words, the economic relationship to the market stands in for nature. In the economic and cultural context of the coalfields, it is easy to oppose environmentalism. Nature, objectified, is reduced to a single tree. The environmental justice movement's holistic view of the human-nature relation contradicts ideals of independent American citizenship and stands in the way of economic progress.

In this context, white coalfield residents protesting against the culture of independent citizenship, private property, and industrial progress often relate themselves to Native Americans. Activists use metaphors of indigeneity to talk about the land differently, as do others who make non-property-based claims to the land. In light of the dominance of corporate landownership, for instance, coalfield residents often assert their nonmarket connections to the land in terms of a long-standing inhabitation that predates the corporate presence. For example, Laura noted that her family had been in the area since "before coal." Denise, an anti-MTR activist, noted about someone who was forced to leave her home due to mining, "To be run off your homeland that, you know, at least ten or eleven generations have grown up on, is ridiculous."

This claim to a nativelike status is entangled in a colonizing logic at the same time as it expresses an aspiration to another way of life. If deployed casually, it erases the original presence of Native Americans in the area and the colonial history of the region. Claims to white victimization can serve as a reassertion of centrality and universality, especially when a near-exclusive focus on white injuries suggests that this suffering is *uniquely* wrong (Mills 1997; Wiegman 1999). From another angle, however, these comparisons also reflect the inherent instability of identity categories. Instead of insisting on a "politics of purity," recognizing the contingency of identity formations opens movements up to emergent coalitions and the possibility of change (A. Smith 2008, 252). Rather than preemptively judging the political correctness of these statements and practices, the fruitful question is where do these comparisons take the coalfield environmental justice movement?

Every single day, coalfield activists stand up against the dismal banality of the dominion of coal. Corrupt politicians, complacent police, and arrogant coal barons are the face of the seemingly interminable reign of King Coal. This dominion is woven into the everyday fabric of American life: air-conditioning, computers, televisions in every room. Disidentifications, enforced by class, gender, and racial formations, materialized in geographies of inequality and the built environment, enable the kinds of NIMBYisms that make coalfield conditions seem like somebody else's problem. Abstractions of space and place in the global commodity system require a formal equality (in the market) that is generated by embodied inequalities of gender and race, class, and citizenship. The sentimental righteousness of the hegemony of coal grimly insists on its conventional stories of modernity, progress, and private property, but the coalfield environmental justice movement mobilizes an alternative morality. Its imagination draws on alternative traditions, including Native American spiritualities and other nonmainstream spiritualities. Linda, an anti-MTR activist, described the goals of the new environmental movement this way: "take the best of the past, things that worked well in everyday life, and the best of the present, and put them together to make a better future." This recalls Deloria's statement about "counting the people who respond to the opportunities of the contemporary world" with a "critical sense of utility" to "serve the people" (Deloria 1999, 240). Linda's discourse was marked by biblical references and Christian overtones but with an environmentalist ethic.[1] She argued, "It's our moral duty to conserve . . . it's a moral issue not to waste the planet for our children, and it's our moral duty to reduce the amount of electricity we use." Referring to an African proverb popularized in the United States by Hillary Clinton, another activist, Gina, emphasized the distinctiveness of this perspective as compared to the dominant neoliberal one: "There's still people around here that believe it takes a village to raise a child." For these activists, what is at stake in the fight over MTR is a particular kind of morality and a particular definition of freedom—all shaped by emphatically nonmarket values.

Anti-MTR activists sometimes explicitly compare their situation to the history of Native American expropriation. For example, a white resident of the town of Blair remarked in a documentary film that now she "know[s] how the Native Americans felt [when they lost their land]."[2] Roger, an anti-MTR activist, told me a story that evoked this connection. A descendant of Chief Cornstalk, a Shawnee leader who died in the Revolutionary War, came to give a blessing to a family cemetery threatened by mining. After giving the blessing, Roger said, the man came up to him and asked, "How does it feel to lose your land?"[3] Similarly, Linda put it this way: "This is a colony. It has been for over 130 years. They call it King Coal and that shows that it's a colony. The people of this area have a lot in common with people of color. Unlike the Indians, who knew what was happening to them, Appalachians don't seem to know what is happening, or they don't want to know."

While this comment reaffirms the "whiteness" of Appalachians, it also reflects Linda's insight that race is part of the problem. At the same time, however, a limiting interval of 130 years disguises the connections between current problems and the area's colonial history. Highlighting these linkages in the context of the fight for environmental justice presents the opportunity for coalfield activists to learn from Native environmental and anticolonial struggles. These metaphors of indigeneity can represent injurious "trophies" of colonialism, but they might also demonstrate the potential for a different relationship to the land and other human communities, an alternative to the one exemplified by MTR. Anti-MTR activists and members of the American Indian Movement have worked together in West Virginia to protect cemeteries from mining and other development, and on other environmental justice issues in the coalfields. The potential for collaboration also exists in commonalities around the issue of protecting children from environmental hazards, the centrality of women in the struggle, and the fight against the divisive strategies of the industry, which support its unsustainable practices.

Emergent environmental coalitions require an ethic of identification, a kind of differential thinking that can decolonize our relationships with each other and the Earth (Sandoval 2000). Anti-MTR discourse often evokes an intersubjective relationship to nonhuman nature that explicitly rejects the hegemonic, objectified one. For example, at a public hearing with the U.S. Office of Surface Mining (OSM) about MTR, in which each speaker had five minutes to talk, Janet Fout, a white activist from the Ohio Valley Environmental Coalition, was granted the floor. Instead of speaking, she started playing a tape of spring peepers and wood thrushes from a mountain forest, saying, "I'm speaking for life; we will all miss the birds and the frogs and the fish." The OSM officials did not allow her to play the tape. One said, "I'm not going to listen to that for five minutes.... What relevance does that have to the stream buffer zone rule? We are not here to hear animal calls or birdcalls." Nonetheless, her act refused the erasure of mountain life that frequently happens in pro-MTR discourse. Because the hearing was about streams and buffer zones, the official asserted that the tape of woodland creatures was not relevant, but the activist was able to make the reductionism of industry and regulatory discourse seem ridiculous (OHVEC 2004).

At the same time, the action of "speaking for . . . the birds and the frogs and the fish" can reflect a reversal of the human/nature binary, a biocentric perspective that privileges nature over humans, replicating the jobs-versus-the-environment aphorism. Guha points out that privileging nature over humans is a form of environmental imperialism that provides a justification for the "continuation of narrow and inequitable conservation practices" that exacerbate human inequalities (Guha 1998, 235). The radical environmentalist view that humans are a cancer on the Earth too often supports a colonizing logic that portrays the least powerful humans as "polluting" and blames them for the problems of poverty and environmental degradation (A. Smith 2005, 63). The coal industry exploits this fear in its workers, the fear that their livelihoods will be bulldozed in the name of an abstract nature. This portrayal of environmentalism resonates with coal miners and their families. After environmental activists from the Rainforest Action Network were arrested for scaling a dragline on Massey's Twilight mountaintop mine, a company spokesman issued a statement:

> It is my understanding that all but one of the fourteen protesters who were arrested [. . .] are residents of states other than West Virginia; such as Maine, Oklahoma, Michigan and Florida. . . . It is clear that these folks are not concerned with the people, the environment or the economy of West Virginia. Their efforts are purely about gaining hype and media attention for their out-of-state funders and patrons. It is time for all West Virginians to stand up and say enough is enough to the protesters. We know our state and our economy and we won't be told what to do. (quoted in Ward 2009c)

This statement is a performative iteration of the relationship between the dancing miner and West Virginia in the Walker Cat ad, which makes claims about the real and plays on people's feelings of powerlessness and marginalization in the national culture. It describes a star-crossed romance between the coal industry and West Virginia in which coalfield anti-MTR activism is rendered invisible. The statement acknowledges the frustration and unhappiness felt by miners and coalfield communities and identifies environmentalist outsiders rather than the coal industry as the problem in an effort to manage people's "ambivalent attachments to the world as such" (Berlant 2008, 21).

Massey's statement attempts to write grassroots Appalachian anti-MTR activism out of what West Virginia means, in a retelling of the jobs-versus-the-environment aphorism that identifies environmentalism as an elitist, unrealistic concern. However, the environmental justice movement offers a potential for overcoming these conventional stories. Environmental justice differs from mainstream environmental conservation in two key ways that signal an emergent paradigm shift in the way environmental problems are framed in American culture. The environmental justice movement started out in communities of color where high rates of illness and disease were the signals that something was wrong (Bullard 1983, 1993; UCC 1987). These community studies demonstrated exactly the kinds of dangers the modern human-nature relation poses to continued human life on Earth, bringing environmental risk home in a way that wilderness conservation does not. Environmental justice starts in the ordinary, everyday experience of the environment, in places not normally recognized in terms of wilderness preservation or beautiful nature. Rather, the inextricable link between human health and happiness and environmental sustainability in everyday life is the focus. Linda expressed the emotional force of an environmental movement founded in social justice and human health, saying of the coal companies, "Damn them to hell for brainwashing our children at the same time as they are poisoning them."

Second, environmental justice movements, including anti-MTR activism, are based in very specific places and human communities. This concreteness offers the potential for overcoming the logic of domination expressed in the human/nature dualism. Rather than expressing abstract environmental values, this grassroots movement is firmly situated in a particular place and in a particular cultural context (Goldman 1996). With its focus on remaining in place, the anti-MTR movement relentlessly centers the connection between environmental and human health; there is nowhere else to go. And because it does not try to generalize its experience or its struggle, this grassroots activism against MTR is less likely to objectify nature or ignore differences among humans.

These differences came out in interesting ways in a conversation I had with a young white activist whose observations suggest some of the differences between what might be called an abstract environmentalism and a localized, place-based one. In his view, true environmentalism means following a particular lifestyle, including a vegetarian diet, and he reported trying to convert other activists to this way of thinking. This belief is also reflected in conflicts that arise between environmentalists and subsistence hunters, in which hunting is assumed to be a destructive activity (A. Smith 2005, 61). The idea that vegetarianism is better for the environment is based on the supposition of an ordinary American supermarket-based diet, and although in many cases this claim holds true it is not generalizable to all contexts. In a rural setting, practices like hunting and meat eating are not necessarily incompatible with environmental flourishing.

The place-based nature of the fight against MTR complicates many assumptions about what environmentalism means. For example, some activists are not against coal mining per se; they advocate a return to underground mining in order to preserve local employment. They see underground mining as a more economically just form of mining (unionized, with strict enforcement of effective safety and environmental regulations), with less devastating effects on the land than MTR.[4] One way they promote this perspective is with a bumper sticker that shows a miner with a lamp on his helmet, crawling on his belly in an underground mine. The sticker reads, "Real coal miners do it deep in the dark." With this slogan that celebrates, through sexually charged imagery, the tough-guy masculinity of underground coal miners over the technocratic modern masculinity of MTR, these activists refute the industry's claim that environmentalists don't care about workers. It also evokes another kind of relationship to the land: in place of the "dead," objectified view of the coal industry, which sees the mountains simply as coal reserves, the slogan evokes an intersubjective relationship between the masculine worker and a feminine nature, a perspective that is historically rooted in European culture, though not currently dominant (Merchant 1980).

This preference for underground mining is not limited to the environmental justice movement. Even Paul, who owns a mining-related business, sees the progress in mining technology as a mixed bag at best. He put it this way: "Your mining complex has changed considerably [through mechanization]. Now, has it changed for the better? . . . I know the company would sit back and say we're making more money. OK, they can get in and mine quicker, but as far as your community, it hasn't helped . . . because you have cut your communities in half. . . . Today everybody is a-moving out to the cities, and . . . then driving an hour back to the jobs, and nobody's looking out for nobody."

The grassroots activists fighting MTR are grappling with the contradictions between conservation and livelihoods in a way that illustrates the potential for place-based coalitions between environmentalists, labor activists, feminists, and antiracists.[5] Epistemologies of social distance, of race, class, and region, identify environmental devastation with less worthy people and require marginalized communities to prove that they are "good" victims who deserve justice. This disidentification is essential to the endurance of a social system based on the exploitation of nature and people. The logic of differential worthiness generates a sense of security, of safety, which enables those who benefit to believe that the costs and benefits allotted by the system make sense. For example, Philip, a white activist, reported nearly losing his temper during a visit to Massachusetts in support of a wind farm off Cape Cod, where irate locals told the coalfield activists to "go back to the hills." This social distance limits people's understanding of the loss and terror caused by environmental devastation and the destruction of communities. Overcoming this distance requires more than just tolerance or an awareness of environmental crises; it requires a realization of the interdependence of distinct social forms and an imagination that can navigate these rhizomic connections.

For instance, consider the frequent comparisons drawn between Appalachia and the Third World. In these comparisons, the experience of people of color—in the global South known patronizingly as the Third World—is used as a metaphor for white suffering. When people exclaim that the coalfields are like the Third World, the metaphor usually contains a suggestion that Americans shouldn't have to live in such conditions. This normalizes both Third World poverty and U.S. privilege. Many coalfield residents take offense to this comparison, hotly replying that West Virginia is in the United States, not the Third World. On the one hand, this comparison is useful because it vividly describes coalfield conditions and the relationship of the region to the rest of the country. On the other hand, it reaffirms the cognitive separation between the coalfields and the rest of the nation. The metaphor of the Third World simultaneously evokes the image of an underdeveloped resource colony and a sense of backwardness and cultural marginality. It carries connotations of what poverty means and how to solve it. The doctrine of modernization as the solution to poverty suggests that places like the coalfields need to progress, to become more like the rest of America. This naturalizes inequalities that are the products of colonizing relationships and valorizes unsustainable practices and social forms. Through the metaphor of the racially, politically, and economically oppressed majority of the planet, the comparison simultaneously indexes an iconic image of suffering in the American imagination and discounts its importance.

At the same time, this decontextualized iteration of the Third World reaffirms the articulation in American culture between progress, wealth, and whiteness. It indicates a lack of development. The phrase is a catchall that naturalizes poverty through cultural connotations such as "crony capitalism," irrational traditions, and parasitic dictatorships. The presence of this comparison in the cultural context of MTR spurs coalfield residents to reaffirm their identity as proper national subjects. Appalachian cultural marginalization constantly puts coalfield residents on the defensive; do they measure up? Are they to blame for their own poverty? Are their communities really worth saving?

Nonetheless, the environmental justice movement in the coalfields is participating in a refocusing of the national environmental movement that, in the increasing urgency of the crisis, is transforming the terms of environmental activism. This new paradigm is increasingly focused on sustainable, democratic community survival and on bridging cultural divides. Recognizing, as one activist put it, paraphrasing Dr. Martin Luther King, that "an injustice anywhere is an injustice everywhere," coalfield activists are fighting their own internalized logic of colonization. During the Mountain Justice summer and spring break training camps antiracism workshops are offered where activists learn about white privilege, histories of racism, and the differences between personal, institutional, and cultural racisms. At the same time, they are challenging mainstream norms of what it means to provide for one's family and community. Appalachian coalfield groups are part of the Power Past Coal coalition, which includes the Indigenous Environmental Network, the Black Mesa Water Coalition, and Dooda (No) Desert Rock. The devastating effects of coal mining and uranium mining on Native American lands are even more invisible in American national culture than the Appalachian case. By working on these causes together, the environmental justice movement uncovers the buried connections between these destructive practices and diverse social forms.

The energy and critical mass of these coalitions move environmental politics beyond representations of essentialized identity categories that promote evaluations of what particular groups *are*, inviting judgments of authenticity, political virtue, or moral worthiness, to questions of what particular groups are *doing*. In 2006, coalfield activists participated in the Environmental Justice for All Tour, which brought together activists fighting mining, toxic waste disposal in minority communities, and workplace safety violations. Coalfield activists participated in PowerShift 2009, where environmental justice advocates and indigenous activists from around the globe affected by coal mining, global warming, and coal-fired power plants joined with youth-led environmental groups to take a stand for green jobs and sustainable energy. The credo of environmental justice—caring for the environment where we live, work, worship, and play—is becoming part of the dominant environmentalist ethos. The ethic of identification that environmental justice embodies refuses to reduce life to an exchange of abstract equivalences. This ethic is rooted in place and community as well as in an imaginative capacity for caring that expands beyond the local.

Recent documentaries on coal and global warming have demonstrated this new vision of connectivity. One features Maria Gunnoe, recipient of a 2009 Goldman Environmental Prize, standing in Times Square, in New York City, where she and several fellow activists have come to speak at the United Nations about mountaintop removal and coalfield public health.[6] As the camera pans across her incredulous face, the wastefulness of Times Square's neon lights becomes impossible to understand. This film places the suffering of Appalachian coalfield residents directly in the context of the "ordinary" excesses of American culture. Describing the impact of coal slurry impoundments on watersheds, a visual in the film demonstrates the spread of this pollution with a black cloud that spreads over the entire eastern United States. This increasingly obvious interconnectivity is making it clear that coal is not just an Appalachian problem.

MTR, in the very excessiveness of its expression of the modern effort to dominate and control nature, also generates a radical critique of this conceptualization of the human relationship to nature. By its violent rupturing of the surface of things, it makes in many ways an ideal site for the study of how modern life is organized. In this book I've explored several aspects of this organization and its far-flung network of articulations. Writing necessarily reduces the complexity of life into containable categories, but the example of MTR highlights the entanglements of divergent fields including cultural processes of subjectification and identification, political economy, human–nonhuman relations, and the production of place.

The production of laboring subjects—the answer to the question of why anyone would choose to be a coal miner—is clearly related to the intersections of class with race, gender, sexuality, and region, among other things.

The choice to become a miner, or the belief in the necessity of MTR, is clearly related to a person's class position, related to living in an economically depressed region, related to having limited choices for a dignified life, and to the separation of life into public and private spheres. However, gender, class, race, and sexuality are not simply systems of oppression but systems of subjectification. Power works in such a way that an "oppressed mountain woman," structured by the dominance of coal mining to be an economically dependent housewife, raising her children in a small mountain community, marginalized and ridiculed in national public discourse, has the potential to become a radical activist, mobilizing her love of a place and her nurturing of a family in an intellectual, moral, and political struggle whose consequences touch all aspects of contemporary life and may help determine the future.

Yet power doesn't simply generate agency in the form of resistance. The question of how MTR is allowed to occur necessitates a reckoning with the many, many ways that everyday life is implicated in the reproduction of economic and environmental inequality. I have attempted to demonstrate some of the ways that personal identity and political subjectivity are wrapped up in the material conditions of life, not only in the conditions of production, in Marxian terms, but also in those elements of everyday life that are often taken for granted. This type of analysis requires moving beyond the dichotomies of public/private and economy/culture. Where a person lives is political, of course, but not without a cultural map that gives the dwelling meaning. To state the obvious, work only occurs within a cultural context. The categories of culture, economy, and politics, which are themselves cultural artifacts, simply make it more difficult to see how deeply cultural an act MTR is.

The conditions of the natural environment are therefore not so much a reflection of patterns of human sociality as they are a coproduction. Indeed, ecofeminists have argued that the dichotomy of humanity and nature is only one of a series of dualisms that structure our unsustainable system (Mies and Shiva 1993; Plumwood 1993; Salleh 1997; R. Stein 2004). People construct themselves through their interaction with the environment, and they do this not only in terms of gender but also through nationality, race, region, and sexuality. Environmental politics therefore exceed the bounds of questions like jobs versus the environment. The environmental justice movement has demonstrated that things like residential segregation are environmental. The race- and gender-segregated job market is environmental. Racial formations are environmental. Constructions of masculinity and femininity are environmental.

The way the environment is entangled in all aspects of life is probably most apparent in the construction of place. Place and place-making practices have obvious environmental dimensions, the most basic of these being the division of city and country. Hence, marginalization is an environmental problem. Appalachia is constructed as a sacrifice zone, a place that can only save itself by destroying itself: inasmuch as it remains unsacrificed, it does not achieve national belonging. The promise of belonging remains unfulfilled, however, because the sacrifice further entrenches the marginalization of the region.

Like other national sacrifice zones, Appalachia has multiple meanings in American national culture. Because of the slippage between Appalachia and "ordinary" America, the region is able to maintain a frontierlike status within the borders of the nation. Inasmuch as it is imagined as an entirely white space, this is a special frontier, a sort of heart of darkness within the heart of modernity. As a perpetual frontier-inside, Appalachia is a site for the ongoing production of American national identity. National dramas are expressed in figurative narratives of gender, race, and class, animating the disidentifications that make environmental destruction possible. The struggles over MTR reflect some of the national struggles over the direction of the American nation, struggles that often hinge on the meaning of concepts like progress, freedom, citizenship, and private property.

Guide to Participants

This alphabetized list of participants is a guide for readers to general information about each of my participants. Not all participants are mentioned by name in the text, but I list them all here to give readers a sense of who was involved with this research. The list includes the pseudonym I assigned according to the individual's culturally legible gender category, along with my estimate of his or her age at the time of interview and the individual's culturally legible race category based on my own understanding of U.S. racial formations. Finally, for context I give the year in which the interview took place.

Alicia, 60, white, resident of Blair, retired store clerk and spouse of
   retired underground miner (2000)
Ben, 50, white, underground miner (2008)
Bill, 50, white, activist (2004)
Chris, 40, white, mining engineer (2000)
Craig, 60, white, retired business owner (2004)
Dave, 30, white, activist (2000)
Denise, 50, white, activist and spouse of underground miner (2008)
Ed, 30, white, underground miner (2008)
Ella, 30, white, spouse of underground miner (2008)
Frances, 40, white, spouse of surface miner (2008)
Fred, 60, black, retired underground miner (2004)
Frieda, 40, white, activist (2004)
Gina, 40, white, activist (2004)
Graham, 50, black, union organizer and former underground
   miner (2008)
Greg, 20, white, underground miner, former resident of Blair (2000)
Gus, 70, white, retired underground miner and resident of Blair
   (2004)
Horace, 70, white, mining-related business owner (2004)

Jack, 60, white, retired underground miner (2000)

Jane, 40, white, activist (2004)

Jay, 50, white, activist (2000)

Jerry, 50, white, educator (2004)

Jesse, 30, white, surface miner (2008)

John, 80, white, state official (2004)

Kevin, 40, white, activist (2004)

Kimberly, 50, black, city official (2004)

Laura, 40, white, educator (2004)

Leann, 30, white, spouse of underground miner (2008)

Linda, 50, white, activist (2004)

Luke, 40, white, underground miner (2008)

Marie, 60, white, retired government employee and spouse of
retired underground miner (2004)

Marjorie, 50, white, cleaner (2004)

Maura, 40, white, activist and spouse of former underground
miner (2008)

Max, 60, white, mining-related business owner (2004)

Meredith, 60, white, spouse of mining-related business owner
(2004)

Nancy, 70, white, widow of underground miner and former
resident of Blair (2000)

Nathan, 50, white, underground miner (2008)

Ned, 50, white, county official (2004)

Nick, 50, white, former miner, financial industry employee (2004)

Nikki, 20, white, activist (2004)

Paul, 60, white, mining-related business owner and former miner
(2004)

Peggy, 50, white, educator (2004)

Pete, 30, white, underground miner (2008)

Peter, 50, white, educator (2004)

Philip, 50, white, activist and former underground miner (2008)

Randall, 60, white, former underground miner (2008)

Randy, 30, white, Department of Environmental Protection officer
(2004)

Renee, 40, white, activist (2000)

Rhonda, 30, white, educator and spouse of underground miner
(2004)

Richard, 70, white, county official (2000)
Rob, 40, white, mining engineering firm employee (2000)
Roger, 50, white, activist (2004)
Sam, 60, white, retired underground miner (2004)
Sandy, 50, white, retired underground miner (2004)
Sarah, 70, white, resident of Blair, spouse of retired underground
miner (2000)
Shawn, 40, white, activist (2004)
Sheri, 40, white, spouse of underground miner (2008)
Stephanie, 60, white, activist (2004)
Steve, 30, white, Department of Environmental Protection officer
(2003)
Suzanne, 50, white, activist (2004)
Tommy, 30, white, underground miner (2008)
Wyatt, 30, white, Department of Environmental Protection officer
(2004)
Zack, 30, white, activist (2008)

Notes

Introduction

1. See the Web sites for Power Shift (http://powershift09.org/), Repower America (http://www.repoweramerica.org/), We Can Solve It (www.wecansolveit.org), and Kilowatt Ours (www.kilowattours.org).

2. The group Mountain Communities for Responsible Energy represents opponents of a proposed wind farm on the highest ridges in Greenbrier County (http://www.wvmcre.org; Giggenbach 2006). This group opposes the wind turbines because they argue that wind farms decrease property values, represent yet another undertaxed exploitation of West Virginian resources for the benefit of others, put the state's tourism industry at risk, are inefficient, and represent significant environmental risk to bats, birds, night views, and natural habitats. The group's Web site offers links to other groups fighting to protect the Appalachian Mountains, including Mountain Justice (http://mountainjusticesummer.org/west_virginia/), which represents groups fighting MTR, such as Coal River Mountain Watch (CRMW). CRMW is currently involved in a campaign to install a wind farm on Coal River Mountain, which will otherwise be the site of a new MTR mine operated by Massey Energy Corporation (http://www.coalriverwind.org/; Ward 2008b).

3. The fetishization of coal is answered by the small model wind turbine that one anti-MTR activist displayed in her home. These small models demonstrate what the activist called the beauty and elegance of this alternative energy source.

4. Goodstein argues that the "trade-off myth" relies on an "economic base model" that is overly simplistic in its understanding of rural economies, that the "direct job-loss numbers" caused by environmental regulation of mining are small, and that this model does not include the possibilities of conversion or restoration work, or what are currently being called "green jobs" (Goodstein 1999, 107–8).

5. "Land-based people" brings to mind "indigeneity," a complicated concept that typically implies a long-term inhabitation of, and cultural identification with, a specific place. Bear in mind that the catchall category "Native American" is itself an artifact of the colonization of the Americas; for Native groups this often means a spiritual and cultural connection with a specific landscape that embodies the group's identity (LaDuke 1999, 2005). In the context of the colonial encounter, the

term connotes tradition as opposed to modernity, and the colonized as opposed to the colonizers. Native Americans and indigeneity have frequently been used as a metonym for the condition of rural Appalachians. For instance, Harry Caudill used the term *indigenous mountaineers* to distinguish between white people who were "native" to the area and more recent European immigrants (Caudill 1962, 125). This use of the term *indigenous* reflects the importance of representations of Native Americans in the U.S. national identity (Kazanjian 2003). These terms are often deployed in reference to Appalachian whites as a sign of primitivism or closeness to nature; additionally, they can represent a claim to territory or they can be used to mark an alternative way of life or a noncommodity orientation to the land (see the Conclusion). These complications reflect the way that marginal whites can perform both a reaffirmation of the racial order and a contestation of that order.

6. This notion elides the fact that MTR reduces the need for labor in mining.

7. International Coal Group (ICG) is the company that owns the Sago mine in Tallmansville, West Virginia, site of the January 2006 disaster that killed twelve men.

8. I am thinking of course of the fetishism of commodities (Marx 1867).

9. Gaventa writes of a kind of split or multiple consciousness among the working class, which he argues prevents the mobilization of the critical consciousness necessary for political action and mires the oppressed in passivity and apathy. According to this model, the effects of power create a weakened consciousness among the oppressed, which renders them more easily dominated. The notion of multiple consciousness has been used since Du Bois first wrote about it to understand the subjectivity of oppressed or marginalized people (Beauvoir 1953; Du Bois 1903). It has not until more recently been applied to unmarked categories of race, gender, and sexuality (Omi and Winant 2009).

10. My analysis proceeds from the assumption that racial categories, like gender and class formations, are the historically mutable products of political and epistemological struggles, and I treat them as real because they have material impacts in the social world. Social categories such as race, gender, and class are powerful cultural organizing principles that affect the way human life and the human–nonhuman relation are organized.

11. My approach differs here from Matt Wray's work on white trash (Wray 2006). He uses a boundary work approach to consider in what sense whiteness operates as a social category apart from racism. I am instead arguing that the stigmatizing of poor whites is actually integral to the maintenance of white racial hegemony over and against people of color (Wellman 2007) at the same time as it interacts productively with other related forms of domination.

12. Half a century ago, Harry Caudill noticed the tendency in eastern Kentucky for people to proclaim that someone they liked was "common," or "not a bit proud" (Caudill 1962, 350). Caudill relates these phrases to the culture of poverty

he saw spreading in the mountains along with the welfare state, and understands them to reflect a lack of dignity, or gumption. Given the local meaning of "pride" however, it is possible to read in these phrases an oblique critique of the rich. Thus, a concern with other people's buying habits might reflect a sense of community in economic hardship.

13. See John Hartigan Jr. on white Appalachians in Detroit (Hartigan 1997a, 1997b, 1999, 2005). Other commonly recognized destinations for out-migration are cities in Ohio or furniture factory locales in the Carolinas. The massive population loss caused by the decline of employment in coal has spawned such popular catchphrases as "Last one out of West Virginia, turn off the light!"

14. The Clean Water Act prohibited mining in and around streams. Valley fills were ruled illegal because they consist of filling streams with mine waste. The Bush EPA changed the classification of overburden from "waste" to "fill material," making valley fills legal (Ward 2002).

15. I have assigned pseudonyms to all participants to protect their confidentiality. I use real names for public figures.

16. See the Appendix for an alphabetical guide to participants.

1. Hillbillies and Coal Miners

1. An example of this type of representation and notation of the frontier is this letter from 1938: "One day we were the happiest people on earth. But like the Indian we are slowly but surely being driven from the homes that we have learned to love. . . . After all I have come to believe that the real old mountaineer is a thing of the past and what will finally take our place God only knows" (William and Wilma Wirt, quoted in Eller 1982, 242).

2. The population of West Virginia is, in fact, mostly white, by about 90 to 95 percent, but its population does include people of all racial categories and has varied historically. Black Appalachians in particular have noted that they are often erased in public discourse that imagines Appalachia as a totally white region. See http://www.wordspy.com/words/Affrilachian.asp for a discussion of the neologism *Affrilachian,* which was coined by a poet to express the experience of African Americans in Appalachia. The Web site quotes the poet Frank X. Walker as having read that Appalachians are defined as "white residents of the mountainous regions of Appalachia."

3. My use of the term *figure* draws on John Hartigan and Donna Haraway (Haraway 1997, 11; Hartigan 2005, 16). A figure in this sense is an affective form of identification that contains a narrative trajectory that can be inhabited and performed creatively in social situations.

4. The tradition of using Native American culture and history as a metonym for American cultural heritage goes back at least to Thomas Jefferson. David

Kazanjian writes that as Jefferson analyzed the contents of a relatively recent Indian burial mound in western Virginia, he simultaneously denied what it represented and claimed this cultural heritage for the United States. "Self-evidently clear to Jefferson, however, is that the Indian's inability to construct a proper, civilized, 'monument' authorizes the treatment of Indian remains as objects of study. . . . The barrows are, to Jefferson, evidence not of the Indians' civil life, but rather of their dead, protocivility. They can thus be considered an assimilable, archeological sign of white settler America's own exceptional history" (Kazanjian 2003, 11).

5. In the United States, political and cultural citizenship has been racialized white (Goldberg 2002; Hong 1999; Mink 1990; Nelson 1998; Valocchi 1994). In the racial formations of European and U.S. imperialism, whiteness was linked to self-governance as well as to modernity and the future, wealth, and development; colonized people were constructed as requiring guidance, their cultures were linked to tradition or backwardness, and their economies were underdeveloped into poverty. In other words, the subject of modernity is (symbolically) white.

6. *Saving Jessica Lynch.* Aired November 9, 2003, by NBC. Directed by Peter Markle and written by John Fasano.

7. *Coal Miner's Daughter.* Released March 7, 1980, by Universal Pictures. Directed by Michael Apted and written by Loretta Lynn and Thomas Rickman.

8. This skit aired on May 8, 2004, season 29, episode 19. Snoop Dogg was the host, and the musical guest was Avril Lavigne. Directed by Beth McCartney and Robert Smigel, written by Doug Abeles and Les Allen.

9. Erickson's award-winning book, *Everything in Its Path* (1976), draws on the work of Harry Caudill and Jack Weller (Caudill 1962; Weller 1965) to describe the social context of the 1972 Buffalo Creek disaster in Logan County. The Buffalo Creek disaster was caused by the failure of a "gob dam." At 8:05 a.m. on February 26, 1972, days of rain finally caused the makeshift dam holding back hundreds of millions of gallons of coal slurry to collapse. The black water flash flooded the communities along Buffalo Creek, killing 125 people and injuring more than a thousand. The flood destroyed over five hundred residences and damaged nearly a thousand more (Erickson 1976). Appalachian studies scholars have pointed out that Erickson's depiction of the Buffalo Creek community in his book relies on the "common sense" representations of Appalachians described here (Ewen and Lewis 1999). In other words, the identity of Appalachians as such is taken as intrinsically informative (Pascale 2007, 37).

10. Observations of the photo exhibit are based on the exhibit at Track 16 Gallery, in Santa Monica, California, on April 22, 2006.

11. Interestingly, for Weller, the poverty of coal miners is explained through their *families'* improper consumption habits; poverty in the coalfields is a failure of responsible familial citizenship (Probyn 1998).

12. Similar photographs could be taken in other very poor places in the United

States—for example, in the rural South or on reservations. However, these images would have very different cultural connotations than those of white Appalachian poverty.

13. *Wrong Turn*. Released May 30, 2003, by Summit Entertainment. Directed by Rob Schmidt and written by Alan B. McElroy.

14. The oddity of this question is the student's relating these real murders to the scenes depicted in *Wrong Turn*.

15. *Deliverance*. Released July 30, 1972, by Warner Brothers. Directed by John Boorman and written by James Dickey; based on the book of the same name by James Dickey. Unlike *Wrong Turn*, *Deliverance* is a well-crafted dramatic thriller that has received critical attention (Armour 1973; Hartigan 2005; R. Wilson 1974). It represents a watershed moment in cultural representations of the mountain South. A few notes of "Dueling Banjos" are enough to conjure up a sense of perversity and sexual threat for many, including people who haven't seen the film. Although the story is set in Georgia, the sense of the backwoods the movie constructs is freely associated with almost any place in Appalachia. The film has been analyzed in terms of its construction of masculinity and national identity. In *Odd Tribes*, Hartigan (2005) analyzes the construction of white identities in *Deliverance*.

16. The internal inconsistency of this narrative, which suggests that civilization emasculates and simultaneously utterly dominates, is arguably part of an ongoing conversation in American culture (Messner 2007).

17. Ironic, because Kentucky is the third largest coal-producing state, after Wyoming and West Virginia, and half of America's electricity comes from coal-burning power plants (see http://www.nma.org/pdf/c_production_state_rank .pdf; Wilen 2008).

18. This is a hardy weed that spreads quickly and takes over fields. The photo's caption simply states "burning wild roses," which leaves room for the interpretation that this is irrational behavior.

19. The film being shown at Track 16 Gallery informed viewers that the Klan members were from out of state and had come to try to recruit from among the hopeless. This information is not in the book.

20. "The Indians Are Coming!" originally aired February 1, 1967, season 5, episode 20, on CBS. Directed by Joseph Depew and written by Paul Henning and Buddy Atkinson.

21. "The South Rises Again" originally aired November 29, 1967, season 6, episode 13, on CBS. Directed by Joseph Depew and written by Paul Henning and Buddy Atkinson.

22. On January 15, 2006, *Larry King Live* on CNN displayed the names and ages of the deceased: Thomas "Tom" Paul Anderson, thirty-nine; Alva Martin "Marty" Bennett, fifty-one; James Arden "Jim" Bennett, sixty-one; Jerry Lee Groves, fifty-six; George Junior Hamner, fifty-four; Terry Michael Helms, fifty; Jesse Logan

Jones, forty-four; David William Lewis, twenty-eight; Martin Toler Jr., fifty-one; Fred Gay "Bear" Ware Jr., fifty-eight; Jackie Lynn Weaver, fifty-one; Marshall Cade Winans, fifty.

23. *Nancy Grace*. CNN. January 4, 2006.

24. Bill O'Reilly. *The O'Reilly Factor*. Fox News. January 4, 2006.

25. Mike Galanos. *Headline Prime*. CNN. January 4, 2006.

26. Dennis O'Dell, interviewed by Anderson Cooper. *Anderson Cooper 360*. CNN. January 4, 2006.

27. A Massey employee I interviewed in 2008 stated that an ICG mine located near his home was highly unsafe and that the company was still cutting corners.

28. Larry King. *Larry King Live*, interview with Tim and Tambra Flint. January 4, 2006.

29. Anderson Cooper reporting. *Nancy Grace*. January 4, 2006.

30. Juliet Huddy and Mike Jerrick. 2006. Interview with Rick McGee. *Dayside*. Fox News. January 4.

31. Fredricka Whitfield. *Memorial for the Sago Miners*. CNN. January 15, 2006.

32. *Anderson Cooper 360*. CNN. January 4, 2006.

33. Ibid. January 6, 2006.

34. Words of Martin Toler Jr. Reported by Soledad O'Brien. *American Morning*. CNN. January 6, 2006.

35. Tim Flint interviewed on *Larry King Live*. CNN. January 4, 2006.

36. Terry Goff interviewed on *Larry King Live*. CNN. January 15, 2006.

37. Anderson Cooper, reporting. *Larry King Live*. CNN January 4, 2006.

38. Julia Bonds and Bill Raney. *Washington Journal*. C-Span. October 17, 2004.

2. Men Moving Mountains

1. Ben's implication that the black miners lacked a work ethic parallels a common middle-class assumption about poor coalfield communities in general.

2. It is possible to earn a middle-class wage from mining in southern West Virginia. Those miners willing to discuss their wages told me they earned around sixty to seventy thousand dollars a year. Many high-skilled mining trades (electrician, continuous miner operator) earn more. However, the hours required to earn these wages are arguably unsustainable in the long term. Jesse, who worked on a Massey MTR mine, reported working sixty-six hours a week for around seventeen dollars an hour.

3. A little over half of the mining families I met were dual-income families.

4. Hegemonic masculinity refers to the ideal type of masculinity in a given place and time. It functions as a disciplinary norm that provides the logical underpinning for the sex/gender system (Connell 1995; Rubin 1997).

5. Though some non-Massey miners talked about eight-hour shifts consisting

of day, evening, or midnight shift, at Massey mines there are frequently only two shifts, each about ten to twelve hours long. On top of these scheduled hours, workers are frequently expected to work overtime (Nyden 2005).

6. In other words, Pete's unexpected need to relieve himself saved him from getting crushed by falling rock. Ella interpreted this as divine intervention.

7. These were probably either the Friends of Coal stickers with various pro-coal slogans like "I love coal" or "WV Coal" or the numerous stickers proclaiming a coal mining identity, such as "I mine coal. You're welcome," or a large decal covering the rearview window of a pickup truck, one of which features the prone, crawling body of an underground miner wearing a headlamp and the text "Six Inches from Hell." The stripes to which Ed refers are the reflective bands of yellow safety tape that mark mining uniforms.

8. Continuous miners are machines that use rotating blades to mine the coal in sections separated by pillars that are left to support the roof. Longwall mining uses a long shearer to mine large sections of the coal seam without roof support. The roof is allowed to fall after the passage of the longwall shearer.

9. This measurement refers to the size of the truck bed.

10. However, Jesse was not satisfied with the wage he earned at Massey.

11. This trip was part of a summer public school enrichment program for kids of diverse ages. I accompanied the group as an observer in order to be able to tour a mountaintop mine.

12. The irony of this statement lies in the fact that currently hundreds of thousands of acres of land have been flattened by MTR. Most of these sites are covered mostly in scrubby grass with a few trees, and only a very small percentage of these sites are actually used for anything. Many are totally isolated and without infrastructure for development.

13. Interestingly, the only other man I interviewed who discussed the state of his marriage was Jesse, a mountaintop miner I talked to by phone. He reported that the main reason he became a mountaintop miner was to earn more money for his wife, which he thought was what she wanted. However, as discussed earlier, he underwent a religious conversion and was very unhappy with his job and his marriage and saw his mistake as having valued earning money over his role as husband and father (which he expressed in the evangelical Christian terms of gender complementarianism).

14. An example of the use of Christian and biblical language in coalfield activism are the dramatic black billboards that proclaim "Stop Destroying My Mountains . . . God" and "In God's hands are the depths of earth and the mountain peaks also belong to God. Psalms 95:4." Unitarian Universalists are also active in the anti-MTR movement.

15. Tragically, this off-the-grid-style independence is threatened by MTR and toxic sludge disposal in the coalfields. Blasting destroys water wells, and toxic waste

poisons these natural water supplies. When this happens, people are forced to fight for city water, which they may not otherwise have desired.

3. The Gendered Politics of Pro–Mountaintop Removal Discourse

1. Recently, however, a court decision from the Fourth Circuit Court of Appeals in Richmond, Virginia, overturned Justice Chambers's previous ruling that called for strict enforcement of the Clean Water Act in issuing permits for MTR (Revkin 2009). Another court decision has challenged the practice of using a streamlined, nationwide permitting process for these mines. According to Justice Goodwin, these "nationwide or 'general' permits are supposed to be used to authorize 'minor activities that are usually not controversial' and that would have only 'minimal cumulative adverse effects on the environment'" (in Ward 2009b). Justice Goodwin ruled that this streamlined process had not taken into account the cumulative effects of MTR. The Obama administration's take on these matters has been equivocal, but the EPA is currently subjecting MTR permits to more scrutiny than did the Bush EPA (Ward 2010).

2. Most of the arguments presented in this chapter are drawn from interviews conducted in 2000, when the mine closure in Blair seemed to threaten the continued operation of the coal industry in West Virginia.

3. Some miners I spoke to had previously worked in other jobs, often for the state. One had been a corrections officer in a prison, another had worked at the state roads department. They each cited the higher pay offered by Massey Energy Company as the reason they left these relatively secure jobs. However, they both lamented exchanging the guaranteed pension from the state for the IRA plan offered by Massey.

4. Chris' judgment on what makes a good worker resonates with the story Philip, a white former miner, told me about his days working at Massey Energy Company. He had heart surgery, and the company urged him to return to work three weeks later, telling him he'd be assigned to answering phones. This is apparently a common practice that reduces the mine's record of time lost to accidents or illness. But after only three days of answering phones, he was ordered back to work underground with his surgery staples still in place.

5. However, this perspective is probably becoming anachronistic; the UMWA is no longer a large bureaucratic union interested in preserving its own power but is becoming an embattled watchdog once again.

6. It is worth noting that both of the men who voiced this opinion were under forty years old. This may have affected their perception of the importance of union programs like the miners' health card and a guaranteed pension.

7. Richard, the county official, contradicted this story of companies willing to help and donate money. He told me about an opportunity that a coal company

had had to help with a water project in a trailer park. He said that if the company had helped, it "would have kept a big lawsuit down, all kinds of stuff. They wouldn't even talk about it. No . . . coal companies are funny."

8. Chris might have been referencing the Coalfield Expressway and the King Coal Expressway, two highway projects that promise to bring economic development into the poorest areas of the coalfields, including a new federal prison planned for McDowell County.

9. Not to mention the global costs of coal-burning power plants.

4. ATVs in Action

1. This kind of use is by definition off trail.

2. But in some eyes this freedom might be restricted to boys—as one white man I interviewed made clear. He suggested that a girl running around with boys would need to be careful not to get a bad reputation.

3. This Appalachian sense of place is beautifully expressed in Catherine Pancake's novel about MTR, *As Strange as This Weather Has Been* (Pancake 2007).

4. This perspective is reflected in one organization's T-shirt, which declares, "Mountaintop removal is hillbilly removal." This also signals the metonymic relationship between white Appalachians and people of color by indexing the fight against "urban renewal."

5. In Gina's case, this was not only due to mining-related environmental hazards and increasing attention to corporate property rights but was probably also related to an atmosphere of hostility responding to her activism.

6. *Christmas in Appalachia*. 1964. CBS News.

7. *A Hidden America: Children of the Mountains*. Aired February 13, 2009, on *20/20*. ABC News. Diane Sawyer reporting.

8. The fact that 80 percent of the users are from out of state is a statement of the program's success in bringing in visitors to West Virginia (West Virginia Legislature's Office of Reference and Information 2005, 6). However, I am interested in how the rationalization of ATV use affects coalfield residents' land use practices.

9. Altina Waller argues that the feud was caused by the capitalist transformation of the area by railroad and coal companies (1988).

5. Coal Heritage/Coal History

1. See http://www.friendsofblairmountain.org.

2. In the first three months of 2006, there were also mining accidents in Canada and Mexico, which are instructive about different national cultures of mining. The Canadian accident occurred in a potash mine. Seventy-two miners were trapped, but there were no deaths for the simple reason that the mine was equipped with

an airtight room stocked with emergency supplies where the miners could go in the event of an explosion. The airtight room kept the miners safe from the poisonous gases released in the explosion and enabled them to wait until the gases had cleared for rescue (Strauss 2006). In the Sago mine disaster, as noted in chapter 1, the miners were equipped with emergency oxygen tanks with enough air only for an hour—the survival of Randal McCloy was termed a miracle for that reason. In Mexico, sixty-five coal miners were trapped by an explosion, and all died (Sherman 2006).

3. National Coal Heritage Area, http://www.cr.nps.gov/heritageareas/areas/ncoal.htm.

4. Coal Heritage Trail, http://www.byways.org/browse/byways/10346/.

5. "First Responders Blocked." February 3, 2006, on WSAZ News Channel 3.

6. How much coal is left is a political question that depends on how the coal is to be accessed and what people are willing to sacrifice for it.

7. A series of signs at the back of the museum lists "Friends of the Coal Heritage Museum." This list includes many local businesses, including some coal operators and other prominent local individuals. Although many coal miners and their families have donated objects to the museum, the major private financial supporters are businesses, not coal miners, and their interests are not necessarily the same as those of the families who have donated objects.

8. Interestingly, this display parallels one in the Oregon Nikkei Legacy Center, where the internment of people of Japanese descent during the Second World War is commemorated, with a similar cot and newspaper, which in that case is dotted with preserved flies evoking the unhygienic conditions in the camp (Anna Tsing, personal communication).

9. Though he cites an interest in "what has happened since" the coal industry's decline in the fifties, this does not seem to be a representative concern of the coal heritage movement.

10. It could coincide generally with the early childhood memories of the baby boom generation, who were born into coal camp families that subsequently left the area. This would explain the preoccupation with Second World War memorabilia, as the baby boomers' parents would be from that generation. And if the miners who are the subjects of coal heritage are also part of the "greatest generation," this puts the coal heritage movement into a national context of nostalgia, as another site of their heroism (Brokaw 1998).

11. Except, of course, for those representations that are specifically concerned with MTR.

12. This encapsulation of coal heritage is an interesting parallel to the phenomenon of Appalachian folk culture. Like the historically distant and timeless "company town," mountain craft culture evokes a homespun, isolated way of life that is supposed to represent the authentic, traditional Appalachian culture. This

culture is represented in and commodified by traditional arts and crafts fairs, places like Tamarack, a West Virginia cultural center and gift shop, and the Foxfire books, which detail how to live off the land in traditional Appalachian ways (Wigginton 1972). Allan Batteau notes how "traditional folk cultures" are generated by an elite's desire for nostalgic commodities representing a retreating past (1990; S. Stewart 1984). Whereas Appalachian folk culture is arguably dominated by feminine practices such as quilting, bonnet wearing, and putting up preserves, coal heritage is an overwhelmingly masculine arena. Perhaps coal heritage represents an answer to the lazy hillbilly stereotypes that accompany the overworked mountain woman.

13. It has also been called the "Red Neck War" because the miners wore red bandannas around their necks to proclaim their radical politics, and some people believe that this is the root of the epithet.

14. Part of this support has been structured by an interdependence of the union with the industry—for one thing, the pensions of retired union miners depend on continued coal production by the industry as a whole. One environmental activist friendly with union miners told me that in this case the union had been intimidated by a small number of unionized MTR miners who vociferously supported the practice.

15. However, this vote has become controversial, with claims of fraud on both sides (Nyden 2010).

16. It is worth noting the SMAP's reference to "visitors from . . . the US and Canada." This statement performs an erasure of Mexico that identifies North America with Anglo-American culture and thereby reasserts the white, mainstream aspirations of the coalfield economic boosters.

17. This is a reference to Denise Giardina, one of West Virginia's most successful contemporary novelists and a former candidate for governor from the Mountain Party.

6. Traces of History

1. The blog "Stuff White People Like" exemplifies this idealization: the "white people" of the blog are uniformly urban, liberal, and sophisticated, albeit ridiculous. See http://stuffwhitepeoplelike.com/.

2. Victoria Johnson points out that the Jeffersonian ideal of a democracy of small, independent producers has multiple valences and, notwithstanding its historical articulation with racial and gendered forms of exclusion, also underlies an underappreciated strain of populism in American political culture (2008).

3. When environmentalists told me there was "plenty" of flat land already, they were usually referring to the hundreds of thousands of undeveloped acres already flattened by MTR. But in this case Shawn was referring to the flat land

that had supported economic activity in Blair prior to the opening of Arch Coal's Daltex mine in 1998, which led to the disintegration of the community, and more recently to the 2004 opening of Arch Coal's Mountaineer II, an underground mine whose sediment pond is now occupying the place that used to be the high school.

4. There are some risks with this kind of postmining land use, however. A federal prison built on a reclaimed mountaintop mine in Kentucky has been dubbed "Sink Sink" because of subsidence (Sloan 2005).

5. In other words, if an environmental protection officer says it's not harming the environment, it's not harming the environment.

6. Recent EPA actions have strengthened these tactics, in particular by emphasizing the effects of water quality on human health.

7. Coalfield communities bear the brunt of the industry's profit imperative in ways that make it clear to them that their well-being is not a priority. Examples include deaths caused by collisions with overweight and speeding coal trucks, increased flooding, and most disturbingly the placement of slurry containment dams (like the ones that collapsed and flooded Buffalo Creek, West Virginia, in 1972, and Harriman, Tennessee, in 2008) immediately above an elementary school, in Marsh Fork, West Virginia (Erikson 1976; Burns 2007; Sledge and Paine 2008; Sludge Safety Project 2009).

8. In the event, Justice Starcher did not run for reelection in 2008.

9. Note here that Blankenship is referencing a normative standard of living with soccer fields, baseball diamonds, and high educational levels.

10. This recalls Barbara Bush's point of view on the condition of people displaced by Hurricane Katrina.

11. Ironically, they gain these rights as long as they do not belong to an exacting homeowners association, as Melanie Dupuis pointed out (personal communication).

Conclusion

1. This use of Christianity is markedly different from the millenarianism of some fundamentalist branches that understand environmental degradation as part of a divine plan.

2. *Razing Appalachia*. Released 2003, by Bullfrog Films, directed by Sasha Waters.

3. Although I originally shared Roger's (apparent) interpretation of this comment, understanding it as a statement of solidarity, I thank Sarita Gaytan for pointing out that it could be interpreted more acerbically (personal communication).

4. The April 5, 2010 disaster at Massey Energy Company's Upper Big Branch mine that killed twenty-nine miners has provided the opportunity for some industry supporters to claim that surface mines are safer than underground mines.

However, what the incident most dramatically illustrated was the danger in working in a nonunion underground mine (Malloy 2010). Some have argued that working on a nonunion *surface mine* is *more* dangerous than working in a unionized underground mine. See http://itsgettinghotinhere.org/2010/04/23/preventing-the-next-mine-disaster-unionize/.

5. The connections between the coalfield situation and that of marginalized people living in resource-rich lands around the country and the world are striking. For instance, Winona LaDuke documents the many Native movements for sustainable energy production, including a proposal for a wind farm on the Lakota Nation's Pine Ridge Reservation in South Dakota (LaDuke 2005, 241–42). Similarly, anti-MTR activists are fighting for a wind farm on Coal River Mountain (Porterfield 2009; Ward 2008b).

6. *Burning the Future: Coal in America.* Released 2007, by Firefly Pix, directed by David Novak. The Goldman Environmental Prize honors six grassroots activists from six continents each year. See http://www.goldmanprize.org.

Bibliography

Adams, Carol J. 2000. *The Sexual Politics of Meat: A Feminist-Vegetarian Critical Theory.* New York: Continuum.

Alarçon, Norma. 1997. "The Theoretical Subject(s) of *This Bridge Called My Back* and Anglo-American Feminism." In *The Second Wave: A Reader in Feminist Theory,* edited by L. Nicholson, 288–99. New York: Routledge.

Anderson, Benedict R. 1983. *Imagined Communities: Reflections on the Origin and Spread of Nationalism.* London: Verso.

Anderson, Justin D. 2008. "Don Blankenship Gets into Tussle with ABC News Reporter." *Charleston Daily Mail,* April 3.

Anglin, Mary K. 1999. "Stories of AIDS in Appalachia." In *Back Talk from Appalachia: Confronting Stereotypes,* edited by D. Billings, G. Norman, and K. Ledford, 267–80. Lexington: University Press of Kentucky.

Anzaldúa, Gloria. 1987. *Borderlands/La Frontera: The New Mestiza.* San Francisco: Spinsters/Aunt Lute Foundation.

Appalachian Land Ownership Task Force. 1983. *Who Owns Appalachia? Land Ownership and Its Impacts.* Lexington: University Press of Kentucky.

Armour, Robert. 1973. "Deliverance: Four Variations on the American Adam." *Literature/Film Quarterly* 3:280–85.

Associated Press. 2004. "T-Shirt Slogan 'Cruel,' W. Va. Governor Says." *Cincinnati Enquirer,* August 26.

Bahnisch, Mark. 2000. "Embodied Work, Divided Labor: Subjectivity and the Scientific Management of the Body in Frederick W. Taylor's 1907 'Lecture on Management.'" *Body & Society* 6 (1): 51–68.

Baker, Houston Jr., 1999. "Critical Memory and the Black Public Sphere." In *Cultural Memory and the Construction of Identity,* edited by L. Weissberg and D. Ben-Amos, 264–96. Detroit: Wayne State University Press.

Barnes, Barbara A. 2010. "Ecoadventures in the American West: Innocence, Conflict, and Nation Making in Emptied Landscapes." In *Toward a Sociology of the Trace,* edited by H. Gray and M. Gómez-Barris. Minneapolis: University of Minnesota Press.

Barrick, Michael. 2009. "Students Guide the Way as University Rids Itself of Coal Stock." The Barrick Report. http://barrickreport.wordpress.com/2009/03/

04/students-guide-the-way-as-university-rids-itself-of-coal-stock/ (accessed March 5, 2009).

Barry, Joyce. 2001. "Mountaineers Are Always Free? An Examination of the Effects of Mountaintop Removal in West Virginia." *Women's Studies Quarterly* 1–2:16–29.

———. 2008. "A Small Group of Thoughtful, Committed Citizens: Women's Activism, Environmental Justice, and the Coal River Mountain Watch." *Environmental Justice* 1 (1): 25–33.

Barthes, Roland. 1976. *Mythologies.* New York: Hill and Wang.

Batteau, Allen. 1990. *The Invention of Appalachia.* Tucson: University of Arizona Press.

Beauvoir, Simone de. 1953. *The Second Sex.* New York: Knopf.

Beck, Ulrich. 1992. *Risk Society: Towards a New Modernity.* London: Sage.

———. 1999. *World Risk Society.* Malden, Mass.: Blackwell.

Beckley Mine. [2007] Youth Museum of Southern West Virginia: Youth Museum Overview http://www.beckleymine.com/ym/ym-overview.cfm (accessed May 12, 2007).

Beckwith, Karen. 2001. "Gender Frames and Collective Action: Configurations of Masculinity in the Pittston Coal Strike." *Politics and Society* 29 (2): 297–330.

Bell, David. 1997. "Anti-Idyll: Rural Horror." In *Contested Countryside Cultures: Otherness, Marginalization, and Rurality,* edited by P. J. Cloke and J. Little, 91–104. London: Routledge.

Berlant, Lauren. 1997. *The Queen of America Goes to Washington City.* Durham, N.C.: Duke University Press.

———. 2008. *The Female Complaint: the Unfinished Business of Sentimentality in American Culture.* Durham, N.C.: Duke University Press.

Bettie, Julie. 2003. *Women without Class: Girls, Race, and Identity.* Berkeley: University of California Press.

Billings, Dwight B., and Katherine M. Blee. 2000. *The Road to Poverty: The Making of Wealth and Hardship in Appalachia.* New York: Cambridge University Press.

Billings, Dwight B., Gurney Norman, and Katherine Ledford. 1999. *Back Talk from Appalachia: Confronting Stereotypes.* Lexington: University Press of Kentucky.

Blee, Kathleen M., and Dwight B. Billings. 1999. "Where 'Bloodshed Is a Pastime': Mountain Feuds and Appalachian Stereotyping." In *Back Talk from Appalachia: Confronting Stereotypes,* edited by D. B. Billings and K. M. Blee, 119–37. Lexington: University Press of Kentucky.

Blomley, Nicholas. 2002. "Mud for the Land." *Public Culture* 14 (3): 557–82.

Bommes, Michael, and Patrick Wright. 1982. "'Charms of Residence': The Public and The Past." In *Making Histories: Studies in History-writing and Politics,* edited by R. Johnson, G. McLennan, B. Schwartz and D. Sutton, 253–301. Minneapolis: University of Minnesota Press.

Bourdieu, Pierre. 1984. *Distinctions: A Social Critique of the Judgement of Taste.* Cambridge, Mass.: Harvard University Press.

————. 1998. *Outline of a Theory of Practice.* New York: Cambridge University Press.

————. 2001. *Masculine Domination.* Stanford, Calif.: StanfordUniversity Press.

Braun, Bruce. 2003. "On the Raggedy Edge of Risk": Articulations of Race and Nature after Biology. In *Race, Nature, and the Politics of Difference,* edited by D. S. Moore, J. Kosek, and A. Pandian, 175–203. Durham, N.C.: Duke University Press.

Brisbin, Richard A. 2002. *A Strike Like No Other Strike: Law and Resistance during the Pittston Coal Strike of 1989–1990.* Baltimore, Md.: Johns Hopkins University Press.

Brokaw, Tom. 1998. *The Greatest Generation.* New York: Random House.

Bullard, Robert D. 1983. "Solid Waste Sites and the Black Houston Community." *Sociological Inquiry* 53:273–88.

————. 1993. *Confronting Environmental Racism: Voices from the Grassroots.* Boston: South End Press.

Burning the Future: Coal in America. 2008. Directed by Davis Novack. Firefly Pix.

Burns, Shirley Stewart. 2007. *Bringing Down the Mountains: The Impact of Mountaintop Removal Surface Coal Mining on Southern West Virginian Communities, 1970–2004.* Morgantown: West Virginia University Press.

Burton, Mark, Michael Hicks, and Calvin Kent. 2000. *Coal Production Forecasts and Economic Impact Simulations in Southern West Virginia: A Special Report to the West Virginia Senate Finance Committee.* Huntington, W.Va.: Center for Business and Economic Research at Marshall University's Lewis College of Business.

Buse, Peter, Ken Hirschkop, Scott McCracken, and Bertrand Taithe. 2005. *Benjamin's Arcades: An Unguided Tour.* Manchester: Manchester University Press.

Butler, Judith. 1990. *Gender Trouble: Feminism and the Subversion of Identity.* New York: Routledge.

————. 1993. *Bodies That Matter: On the Discursive Limits of Sex.* New York: Routledge.

Byrnes, Nanette, and Adam Aston. 2007. "Coal? Yes, Coal: Never Mind Global Warming." Peabody Energy CEO Gregory Boyce Is Betting Big on the Dirty Fuel's Durability. *BusinessWeek,* May 7. http://www.businessweek.com/print/magazine/content/07_19/b4033075.htm?chan=gl (accessed February 3, 2009).

Caldwell, Alison. 2004. "Town Caught up in Iraqi Prisoner Abuse Controversy." *The World Today: ABC News,* May 7. Transcript of broadcast. http://www.abc.net.au/worldtoday/content/2004/s1103623.htm (accessed August 4, 2006).

Carby, Hazel V. 1982. "White Woman Listen! Black Feminism and the Boundaries of Sisterhood." In *The Empire Strikes Back: Race and Racism in 70's Britain,* edited by P. Gilroy and H. V. Carby, 221–234. London: Routledge.

Caudill, Harry. 1962. *Night Comes to the Cumberlands.* Boston: Little, Brown and Co.

Chafin, Barbara. 2005. "Mountaintop Removal in Mingo County—Without a Permit!" *Winds of Change: the Newsletter of the Ohio Valley Environmental Coalition,* May. http://www.ohvec.org/newsletters/woc_2005_05/article_18.html (accessed August 21, 2005).

Clines, Francis X. 1999a. "Last Stand in Defense of a Hollow's History." *New York Times,* November 7.

———. 1999b. "Senator Leads Move to Block Ruling on Strip Mines. " *New York Times,* October 30.

———. 1999c. "With 500 Miners as a Chorus, Byrd Attacks Court Ruling. " *New York Times,* November 7.

Cohen, Phil. 1997. "Laboring under Whiteness. " In *Displacing Whiteness: Essays in Social and Cultural Criticism,* edited by R. Frankenberg, 244–282. Durham, N.C.: Duke University Press.

Cohn, Carol. 1993. "Wars, Wimps and Women: Talking Gender and Thinking War." In *Gendering War Talk,* edited by M. Cooke and A. Woollacott, 227–47. Princeton, N.J.: Princeton University Press.

Colford, Paul D., and Corky Siemaszko. 2003. "Fiends Raped Jessica." *New York Daily News,* November 6.

Collier, Richard. 1995. *Masculinity, Law, and the Family.* New York: Routledge.

Connell, R. W. 1993. "The Big Picture: Masculinities in Recent World History." *Theory and Society* 22 (5): 597–623.

———. 1995. *Masculinities.* Berkeley: University of California Press.

Connerton, Paul. 2006. "Cultural Memory." In *Handbook of Material Culture,* edited by C. Tilley, W. Keane, S. Kuchler, M. Rowlands, and P. Spyer, 315–24. Thousand Oaks, Calif.: Sage Publications.

Coombe, Rosemary J. 2001. "Sports Trademarks and Somatic Politics: Locating the Law in Critical Cultural Studies." In *Between Law and Culture: Relocating Legal Studies,* edited by D. T. Goldberg, M. Musheno and L. C. Bower, 22–49. Minneapolis: University of Minnesota Press.

Coombes, Annie E. 2003. *History after Apartheid: Visual Culture and Public Memory in a Democratic South Africa.* Durham, N.C.: Duke University Press.

Corbin, David Alan. 1981. *Life, Work, and Rebellion in the Coal Fields: The Southern West Virginia Miners 1880–1922.* Urbana: University of Illinois Press.

Craig, Lyn. 2006. "Does Father Care Mean Fathers Share? A Comparison of How Mothers and Fathers in Intact Families Spend Time with Children." *Gender and Society* 20 (2): 259–81.

Cronon, W. 1983. *Changes in the Land: Indians, Colonists and the Ecology of New England.* New York: Hill and Wang.

———. 1991. *Nature's Metropolis: Chicago and the Great West.* New York: W. W. Norton & Co.

Cunningham, Dan. 1946. "The Horrible Butcheries of West Virginia." *West Virginia History Journal*, 46: 25–44.

Cuomo, Chris. 2005. "Ethics and the Eco/Feminist Self." In *Environmental Philosophy: from Animal Rights to Radical Ecology*, edited by M. E. Zimmerman, 194–207. Upper Saddle River, N.J.: Pearson Prentice Hall.

Curtin, Deane W. 2005. *Environmental Ethics for a Postcolonial World*. London: Rowan & Littlefield.

Dao, James. 2004. "T-Shirt Slight Has West Virginia in Arms." *New York Times*, March 23.

———. 2005. "A New Campaign to Preserve an Old Mining Battlefield." *New York Times*, May 15.

Deloria, Vine, Jr. 1999. *Spirit and Reason: The Vine Deloria, Jr., Reader*. Golden, Colo.: Fulcrum Publishing.

Derickson, Alan. 1998. *Black Lung: Anatomy of a Public Heath Disaster*. Ithaca, N.Y.: Cornell University Press.

Derrida, Jacques, and Peter Eisenman. 1997. *Chora L Works*. Edited by J. Kipnis and T. Leeser. New York: Monacelli Press.

Descola, Philippe. 1994. *In the Society of Nature: A Native Ecology in Amazonia*. New York: Cambridge University Press.

Di Leonardo, Micaela. 1998. *Exotics at Home: Anthroplogies, Others, American Modernity*. Chicago: University of Chicago Press.

Douglas, Mary. 2002 (1966). *Purity and Danger*. New York: Routledge Classics.

Du Bois, W. E. B. 1903 (1982). *The Souls of Black Folks*. New York: New American Library.

DuPuis, E. Melanie. 2002. *Nature's Perfect Food: How Milk Became America's Drink*. New York: New York University Press.

Dyer, Richard. 1997. *White*. New York: Routledge.

———. 2000. "The Role of Stereotypes." In *Media Studies: A Reader*, edited by P. Marris and S. Thornham, 245–51. New York: New York University Press.

Eichstedt, Jennifer, and Stephen Small. 2002. *Representations of Slavery: Race and Ideology in Southern Plantation Museums*. Washington, D.C.: Smithsonian Institution Press.

Eller, Ronald D. 1982. *Miners, Millhands, and Mountaineers: Industrialization of the Appalachian South, 1880–1930*. Knoxville: University of Tennessee Press.

ePlans.com. [2006]. New American House Plans. http://www.eplans.com/new-american-house-plans/new-american.hwx (accessed August 9, 2006).

Erikson, Kai. 1976. *Everything in Its Path: Destruction of Community in the Buffalo Creek Flood*. New York: Simon & Schuster.

Ewen, Lynda Ann, and Julie Lewis. 1999. "Revisiting Buffalo Creek and Everything in Its Path: Deconstructing an Outsider's Stereotypes." *Appalachian Journal: A Regional Studies Review* 27 (1): 22–45.

Fine, Michelle, Lois Weis, Judi Addelston, and Julia Marusza. 1997. "(In)Secure Times: Constructing White Working-Class Masculinities in the Late 20th Century." *Gender and Society* 11 (1): 52–68.

Fisher, Steven L. 1990. "National Economic Renewal Programs and Their Impact on Appalachia and the South." In *Communities in Economic Crisis: Appalachia and the South,* edited by J. Gaventa, B. E. Smith, and A. Willingham, 263–78. Philadelphia: Temple University Press.

Fleming, Caleb. 2009. "Coal Slurry Research May Help W. Va. Community." *Virginia Tech Collegiate Times,* February 17.

Flores, William V. 1997. "*Mujeres en Huelga*: Cultural Citizenship and Gender Empowerment in a Cannery Strike." In *Latino Cultural Citizenship: Claiming Identity, Space, and Rights,* edited by W. V. Flores and R. Benmayor, 210–54. Boston: Beacon Press.

Foley, Ken. 1997. *The White Scourge: Mexicans, Blacks, and Poor Whites in Texas Cotton Culture.* Berkeley: University of California Press.

Foster, Stephen William. 1988. *The Past Is Another Country: Representation, Historical Consciousness, and Resistance in the Blue Ridge.* Berkeley: University of California Press.

Foucault, Michel. 1975 (1995). *Discipline and Punish: The Birth of the Prison.* New York: Vintage Books.

———. 1976 (1990). *The History of Sexuality,* Vol. 1. New York: Vintage.

Fox, Julia. 1999. "Mountaintop Removal in West Virginia: An Environmental Sacrifice Zone." *Organization & Environment* 12 (2): 163–83.

Frankenberg, Ruth. 1993. *White Women, Race Matters: the Social Construction of Whiteness.* Minneapolis: University of Minnesota Press.

———. 1997. "Introduction: Local Whitenesses, Locating Whiteness." In *Displacing Whiteness: Essays in Social and Cultural Criticism,* edited by R. Frankenberg, 1–33. Durham, N.C.: Duke University Press.

Fraser, Nancy. 1995. "What's Critical about Critical Theory?" In *Feminists Read Habermas: Gendering the Subject of Discourse,* edited by J. Meehan, 21–56. New York: Routledge.

Freese, Barbara. 2003. *Coal: A Human History.* Cambridge, Mass.: Perseus Publishing.

Frost, William Goodell. 1899 (1995). "Our Contemporary Ancestors in the Southern Mountains." In *Appalachian Images in Folk and Popular Culture,* 91–106. Knoxville: University of Tennessee Press.

Gallagher, Sally K. 1999. "Symbolic Traditionalism and Pragmatic Egalitarianism: Contemporary Evangelicals, Families, and Gender." *Gender and Society* 13 (2): 211–33.

Garlick, Steve. 2003. "What Is a Man? Heterosexuality and the Technology of Masculinity." *Men and Masculinities* 6 (2): 156–72.

Garroutte, Eva Marie. 2003. *Real Indians: Identity and the Survival of Native America*. Berkeley: University of California Press.

Gaventa, John. 1980. *Power and Powerlessness: Quiescence and Rebellion in an Appalachian Valley*. Urbana: University of Illinois Press.

Gaventa, John, Barbara Ellen Smith, and Alex Willingham. 1990. *Communities in Economic Crisis: Appalachia and the South*. Philadelphia: Temple University Press.

Gaytan, Marie Sarita. 2010. "Drinking the Nation and Making Masculinity: Tequila, the Revolution, and Mexican Identity." In *Toward a Sociology of the Trace*, edited by H. Gray and M. Gómez-Barris. Minneapolis: University of Minnesota Press.

Giardina, Denise. 1994. *The Unquiet Earth*. New York: Ballantine Books.

Gibbs, Nancy. 2003. "The Private Jessica Lynch." *Time*, November 17.

Gibson-Graham, J. K. 1996. *The End of Capitalism (as We Knew It)*. Cambridge, Mass.: Blackwell.

Giddens, Anthony. 1984. *The Constitution of Society*. Berkeley: University of Californa Press.

———. 1990. *The Consequences of Modernity*. Stanford, Calif.: Stanford University Press.

Giggenbach, Christian. 2006. "PSC OKs Greenbrier County Wind Farm." *Beckley (W.Va.) Register–Herald*, August 29.

Goldberg, David Theo. 1993. *Racist Culture: Philosophy and the Politics of Meaning*. Cambridge, Mass.: Blackwell.

———. 2002. *The Racial State*. Cambridge, Mass.: Blackwell.

Goldman, Benjamin A. 1996. "What Is the Future of Environmental Justice?" *Antipode* 28 (2): 122–41.

Goodell, Jeff. 2006. *Big Coal: The Dirty Secret behind America's Energy Future*. New York: Houghton Mifflin Company.

Goodstein, Eban S. 1999. *The Trade-off Myth: Fact and Fiction about Jobs and the Environment*. Washington, D.C.: Island Press.

Gordon, Avery. 1997. *Ghostly Matters: Haunting and the Sociological Imagination*. Minneapolis: University of Minnesota Press.

———. 1999. "Globalism and the Prison Industrial Complex: An Interview with Angela Davis." *Race & Class* 40 (2–3): 145–57.

Gordon, Linda. 1990. *Women, the State and Welfare*. Madison: University of Wisconsin Press.

———. 1994. *Pitied but Not Entitled: Single Mothers and the History of Welfare, 1890–1935*. New York: Free Press.

Gottdiener, Mark. 2001. *The Theming of America: American Dreams, Media Fantasies, and Themed Environments*. New York: Blackwell.

Gottlieb, Robert. 1993. *Forcing the Spring: the Transformation of the American Environmental Movement*. Washington, D.C.: Island Press.

Graham, Allison. 2001. *Framing the South: Hollywood, Television and Race during the Civil Rights Struggle.* Baltimore: Johns Hopkins University Press.

Gray, Herman. 1995. *Watching Race: Television and the Struggle for Blackness.* Minneapolis: University of Minnesota Press.

Greene, Janet W. 1990. "Strategies for Survivial: Women's Work in the Southern West Virginia Coal Camps." *West Virginia History* 49:37–54.

Guha, Ramachandra. 1998. "Radical American Environmentalism and Wilderness Preservation: a Third World Critique." In *The Great New Wilderness Debate,* edited by J. B. Callicott and M. P. Nelson, 231–45. Athens: University of Georgia Press.

Haas, Johanna Marie. 2008. "Law and Property in the Mountains: A Political Economy of Resource Land in the Appalachian Coalfields." PhD diss., geography, Ohio State University.

Halberstram, Judith. 2006. *Female Masculinity.* Durham, N.C.: Duke University Press.

Hall, Stuart. 1988. "The Toad in the Garden: Thatcherism among the Theorists." In *Marxism and the Interpretation of Culture,* edited by C. Nelson and L. Grossberg. Chicago: University of Illinois Press.

———. 1996. *Race: The Floating Signifier.* Directed by Sut Jhally. Video. Northampton, Mass.: Media Education Foundation.

———. 2000. "Racist Ideologies and the Media." In *Media Studies: A Reader,* edited by P. Marris and S. Thornham, 271–82. New York: New York University Press.

Halsall, Paul. 1998. "Modern History Sourcebook: Women Miners in the English Coal Pits." Fordham University. http://www.fordham.edu/halsall/mod/1842womenminers.html (accessed May 29, 2001).

Haney-López, Ian F. 1996. *White by Law: The Legal Construction of Race.* New York: New York University Press.

Haraway, Donna Jeanne. 1997. *Modest Witness@Second Millennium.FemaleMan Meets OncoMouse: Feminism and Technoscience.* New York: Routledge.

Harris, Cheryl. 1993. "Whiteness as Property." *Harvard Law Review* 106:1707–91.

Harrison, Harry. 1946. *Their Ancient Grudge.* Indianapolis, Ind.: Bobbs-Merrill.

Hartigan, John, Jr. 1997a. "Locating White Detroit." In *Displacing Whiteness: Essays in Social and Cultural Criticism,* edited by R. Frankenberg, 180–213. Durham, N.C.: Duke University Press.

———. 1997b. "Name Calling: Objectifying 'Poor Whites' and 'White Trash' in Detroit." In *White Trash: Race and Class in America,* edited by M. Wray and A. Newitz, 41–56. New York: Routledge.

———. 1999. *Racial Situations: Class Predicaments of Whiteness in Detroit.* Princeton, N.J.: Princeton University Press.

———. 2005. *Odd Tribes: Toward a Cultural Analysis of White People.* Durham, N.C.: Duke University Press.

Hartman, Saidiya. 1999. "Seduction and the Ruses of Power." In *Between Woman and Nation: Nationalisms, Transnational Feminisms, and the State,* edited by C. Kaplan, N. Alarcón, and M. Moallem, 111–41. Durham, N.C.: Duke University Press.

Hartmann, Heidi. 1976. "Capitalism, Patriarchy and Job Segregation by Sex." *Signs* 1 (3): 137–69.

Harvey, David. 1990a. "Flexible Accumulation through Urbanization: Reflections on 'Post-Modernism' in the American City." *Perspecta* 26:251–72.

———. 1990b. *The Condition of Postmodernity.* Oxford, Mass.: Blackwell.

Heneghan, Bridget T. 2003. *Whitewashing America: Material Culture and Race in the Antebellum Imagination.* Jackson: University Press of Mississippi.

Hilden, Patricia Penn. 1993. *Women, Work, and Politics: Belgium, 1830–1914.* Oxford, England: Clarendon Press.

Hill, Herbert. 1988. "Myth-Making as Labor History: Herbert Gutman and the United Mine Workers of America." *Politics, Culture, and Society* 2 (2): 132–200.

Hill, Joan G. 2007. Review of *Coal Hollow: Photographs and Oral Histories,* by Ken Light and Melanie Light. *Labor Studies Journal,* 31 (4): 87–88.

Hill Collins, Patricia. 2000. *Black Feminist Thought: Knowledge, Consciousness, and the Politics of Empowerment.* New York: Routledge.

Hilson, Gavin M. 2003. *The Socio-Economic Impacts of Artisanal and Small-Scale Mining in Developing Countries.* Exton, Penn.: A. A. Balkerma.

Hinton, Jennifer J., Marcello M. Viega, and Christian Beinhoff. 2003. "Women and Artisanal Mining: Gender Roles and the Road Ahead." In *The Socio-Economic Impacts of Artisanal and Small-Scale Mining in Developing Countries,* edited by G. M. Hilson, 161–203. Exton, Penn.: A. A. Balkerma.

Hirsh, Michael, John Barry, and Babak Dehghanpisheh. 2004. "'Hillbilly Armor': Defense Sees It's Fallen Short in Securing the Troops. The Grunts Already Knew." *Newsweek,* December 20.

Hoggart, Richard, and Raymond Williams. 1960. "Working Class Attitudes." *New Left Review* 1:26–30.

Hollywood, Amy. 2002. "Performativity, Citationality, and Ritualization." *History of Religions* 42 (2): 93–115.

Hong, Grace Kyungwon. 1999. "'Something Forgotten Which Should Have Been Remembered': Private Property and Cross-Racial Solidarity in the Work of Hisaye Yamamoto." *American Literature* 71 (2): 291–310.

Horowitz, Roger. 2001. *Boys and Their Toys: Masculinity, Class, and Technology.* New York: Routledge.

Hurtado, Aida. 1996. *The Color of Privilege: Three Blasphemies on Race and Feminism.* Ann Arbor: University of Michigan Press.

Ignatiev, Noel. 1995. *How the Irish Became White*. New York: Routledge.

Ingold, Tim. 2000. *The Perception of the Environment: Essays on Livelihood, Dwelling and Skill*. London: Routledge.

———. 2004. "Culture on the Ground: The World Perceived through the Feet." *Journal of Material Culture* 9:315–40.

Inscoe, John C. 1999. "The Racial 'Innocence' of Appalachia: William Faulkner and the Mountain South." In *Backtalk from Appalachia: Confronting Stereotypes*, edited by D. Billings, G. Norman, and K. Ledford, 85–97. Lexington: University Press of Kentucky.

International Coal Group, Inc. 2008. *Annual Report. U.S. Securities and Exchange Commission*. http://idea.sec.gov/Archives/edgar/data/1320934/000119312 508043940/d10k.htm (accessed February 11, 2009).

Internet Movie Database. [2006]. "Trivia for 'The Beverly Hillbillies.'" http://www .imdb.com/title/tt0055662/trivia (accessed April 16, 2006).

Jameson, Fredric. 1984. "Postmodernism, or The Cultural Logic of Late Capitalism." *New Left Review* 146:53–92.

———. 1991. "Cognitive Mapping." In *Marxism and the Interpretation of Culture*. Edited by C. Nelson and L. Grossberg, 347–60. Indianapolis, Ind.: University of Illinois Press.

Janofsky, Michael. 1998. "As Hills Fill Hollows, Some West Virginians Are Fighting King Coal." *New York Times*, May 7.

Jefferson, Thomas. 1794 (1982). *Notes on the State of Virginia*. Edited by W. Peden. Chapel Hill: University of North Carolina Press.

Johnson, Richard, Graham Dawson, and the Popular Memory Group. 1982. "Popular Memory: Theory, Politics, Method." In *Making Histories: Studies in History Writing and Politics*, edited by R. Johnson, G. McLennan, B. Schwartz, and D. Sutton, 205–52. Minneapolis: University of Minnesota Press.

Johnson, Victoria. 2008. *How Many Machine Guns Does It Take to Cook One Meal? The Seattle and San Francisco General Strikes*. Seattle: University of Washington Press.

Kane, Emily. 2006. "'No Way My Boys Are Going to Be Like That! Parents' Responses to Children's Gender Nonconformity." *Gender and Society* 20 (2): 149–76.

Kazanjian, David. 2003. *The Colonizing Trick: National Culture and Imperial Citizenship in Early America*. Minneapolis: University of Minnesota Press.

Kimmel, Michael. 1994. "Masculinity as Homophobia: Fear, Shame, and Silence in the Construction of Gender Identity." In *Theorizing Masculinities*, edited by H. Brod and M. Kaufman, 119–41. Thousand Oaks, Calif.: Sage Publications.

King, Byron W. 2009. "How Much Coal Is Out There?" *Energy Bulletin*. Agora Financial, March 2. http://www.energybulletin.net/node/48240 (accessed March 4, 2009).

Klotter, James. 1980. "The Black South and White Appalachia." *Journal of American History* 66 (4): 832–49.

Kolbert, Elizabeth. 2009. "The Sixth Extinction. *New Yorker,* May 25.

Koma, Victor. [2006]. "'The Beverly Hillbillies' and Buick?" PrewarBuick.com. http://www.prewarbuick.com/features/jed_clampetts_buick (accessed April 16, 2006).

Kondo, Dorinne. 1997. *About Face: Performing Race in Fashion and Theater.* New York: Routledge.

Kristeva, Julia. 1982. *Powers of Horror: An Essay on Abjection.* New York: Columbia University Press.

Kuletz, Valerie L. 1998. *The Tainted Desert: Environmental and Social Ruin in the American West.* New York: Routledge.

LaDuke, Winona. 1999. *All Our Relations: Native Struggles for Land and Life.* Cambridge, Mass.: South End Press.

———. 2005. *Recovering the Sacred: The Power of Naming and Claiming.* Cambridge, Mass.: South End Press.

Latour, Bruno. 1993. *We Have Never Been Modern.* Cambridge, Mass.: Harvard University Press.

Lefebvre, Henri. 1991. *The Production of Space.* Cambridge, Mass.: Blackwell.

Levin, Amy. 2007. "Introduction to Part III: Nostalgia as Epistemology." In *Defining Memory: Local Museums and the Construction of History in America's Changing Communities,* edited by A. Levin, 93–96. Lanham, Md.: Altamira Press.

Lewis, Helen Matthews, Linda Johnson, and Donald Askins. 1978. *Colonialism in Modern America: the Appalachian Case.* Boone, N.C.: Appalachian Consortium Press.

Light, Ken, and Melanie Light. 2006. *Coal Hollow: Photographs and Oral Histories.* Berkeley: University of California Press.

Lipsitz, George. 1998. *The Possessive Investment in Whiteness: How White People Profit from Identity Politics.* Philadelphia: Temple University Press.

Liptak, Adam. 2008. "Motion Ties W. Virginia Justice to Coal Executive." *New York Times,* January 15.

———. 2009. "Justices Hear Arguments on Court-Money Nexus." *New York Times,* March 3.

Locke, John. 1690 (1998). "Two Treatises on Government: Second Treatise." In *Social and Political Theory: Classical Readings,* edited by M. Kimmel and C. Stephen, 25–31. Boston: Allyn and Bacon.

Loeb, Penny. 2007. *Moving Mountains: How One Woman and Her Community Won Justice from Big Coal.* Lexington: University Press of Kentucky.

Lohan, Maria, and Wendy Faulkner. 2004. "Masculinities and Technologies: Some Introductory Remarks." *Men and Masculinities* 6(4): 319–29.

Lorber, Judith. 1993. "Believing Is Seeing: Biology as Ideology." *Gender and Society* 7 (4): 568–81.

Lucal, Betsy. 1999. "What It Means to Be Gendered Me." *Gender and Society* 13 (6): 781–97.

Lupton, Ben. 2006. "Explaining Men's Entry into Female-Concentrated Occupations: Issues of Masculinity and Social Class." *Gender, Work, and Organization* 13 (2): 103–28.

Maggard, Sally Ward. 1999. "Gender, Race, and Place: Confounding Labor Activism in Central Appalachia." In *Neither Separate nor Equal: Women, Race, and Class in the South,* edited by B. E. Smith, 185–206. Philadelphia: Temple University Press.

Malloy, Daniel. 2010. "Are Union Mines Safer? W. Va. Tragedy at Nonunion Site Rekindles Debate." *Pittsburgh Post-Gazette,* April 18.

Marx, Karl. 1867 (1978). "Capital, Volume One (Excerpts)." In *The Marx-Engels Reader,* edited by R. C. Tucker, 294–438. New York: W. W. Norton.

———. 1848 (1978). "The Manifesto of the Communist Party." In *The Marx-Engels Reader,* edited by R. C. Tucker, 469–500. New York: W. W. Norton.

Massey, Doreen. 1993. "Power-Geometry and a Progressive Sense of Place." In *Mapping the Futures: Local Cultures, Global Change,* edited by J. Bird, B. Curtis, T. Putnam, G. Robertson, and L. Tickner, 59–69. London: Routledge.

———. 1994. *Space, Place, and Gender.* Minneapolis: University of Minnesota Press.

Massey Energy Company. 2008. *Annual Report.* U.S. Securities and Exchange Commission. http://idea/sec.gov/Archives/edgar/data/37748/000003774 808000012/form10k123108.htm (accessed February 11, 2009).

McClintock, Anne. 1995. *Imperial Leather: Race, Gender, and Sexuality in the Colonial Context.* New York: Routledge.

McCloy, Randal. 2006. "Letter to the Families and Loved Ones of My Co-Workers, Victims of the Sago Mine Disaster." *Charleston Gazette,* April 28.

McNeil, W. K. 1995. *Appalachian Images in Folk and Popular Culture.* Knoxville: University of Tennessee Press.

McNeill, Tanya. 2010. "A Nation of Families: The Codification and (Be)longings of Heteropatriarchy." In *Toward a Sociology of the Trace,* edited by H. Gray and M. Gómez-Barris. Minneapolis: University of Minnesota Press.

Meador, Michael. 1981. "The Redneck War of 1921: The Miners' March and the Battle of Blair Mountain." *Goldenseal,* April–June.

Merchant, Carolyn. 1980. *The Death of Nature.* New York: HarperCollins.

———. 1995. "Reinventing Eden: Western Culture as Recovery Narrative." In *Uncommon Ground: Toward Reinventing Nature,* edited by W. Cronon, 132–70. New York: Norton.

Messner, Michael A. 2000. "Barbie Girls versus Sea Monsters: Children Constructing Gender." *Gender and Society* 14 (6): 765–84.

————. 2007. "The Masculinity of the Governator: Muscle and Compassion in American Politics." *Gender and Society*. 21 (4): 461–80.

Mies, Maria, and Vandana Shiva. 1993. *Ecofeminism*. New Delhi: Kali for Women.

Miller, Tom D. 1974. "Who Owns West Virginia?" *Huntington Herald-Advertiser and Herald-Dispatch*, special section.

Mills, Charles W. 1997. *The Racial Contract*. Ithaca, N.Y.: Cornell University Press.

Mink, Gwendolyn. 1990. "The Lady and the Tramp: Gender, Race and the Origins of the American Welfare State." In *Women, the State, and Welfare*, edited by L. Gordon, 92–122. Madison: University of Wisconsin Press.

Mintz, Sidney. 1986. *Sweetness and Power: The Place of Sugar in Modern History*. New York: Penguin Books.

Mohanty, Chandra Talpade. 1997. "Women Workers and Capitalist Scripts: Ideologies of Domination, Common Interests, and the Politics of Solidarity." In *Feminist Genealogies, Colonial Legacies, Democratic Futures*, edited by M. J. Alexander and C. Mohanty, 3–29. New York: Routledge.

Montrie, Chad. 2003. *To Save the Land and People: A History of Opposition to Surface Coal Mining in Appalachia*. Chapel Hill: University of North Carolina Press.

Moore, Donald S., Jake Kosek, and Anand Pandian. 2003. *Race, Nature, and the Politics of Difference*. Durham, N.C.: Duke University Press.

Moore, Marat. 1996. *Women in the Mines: Stories of Life and Work, Twayne's Oral History Series*. New York: Twayne Publishers.

Moskowitz, Marina. 2004. *Standard of Living: the Measure of the Middle Class in Modern America*. Baltimore: Johns Hopkins University Press.

Mukherjee, Roopali. 2005. *The Racial Order of Things: Cultural Imaginaries of the Post-Soul Era*. Minneapolis: University of Minnesota Press.

Myers, Jeffrey. 2005. *Converging Stories: Race, Ecology, and Environmental Justice in American Literature*. Athens: University of Georgia Press.

Naono, Akiko. 2010. "Producing Sacrificed Subjects for the Nation: Japan's War-Related Redress Policy and the Endurance Doctrine." In *Toward a Sociology of the Trace*, edited by H. Gray and M. Gómez-Barris. Minneapolis: University of Minnesota Press.

NCHA (National Coal Heritage Area). [2006]. "The National Coal Heritage Area: Welcome to the National Coal Heritage Area!" http://www.coalheritage.org/moreabout.html (accessed February 2, 2006).

Nelson, Dana D. 1998. *National Manhood: Capitalist Citizenship and the Imagined Fraternity of White Men*. Durham, N.C.: Duke University Press.

Newitz, Annalee, and Matt Wray. 1997. "What Is 'White Trash'? Stereotypes and Economic Conditions of Poor Whites in the United States." In *Whiteness: A Critical Reader*, edited by M. Hill, 168–84. New York: New York University Press.

Norris, Randall, and Jean-Philippe Cyprès. 1996. *Women of Coal*. Lexington: University Press of Kentucky.

NRHP (National Register of Historic Places). [2006]. "The National Register of Historic Places: What Are the Results of Listing?" http://www.cr.nps.gov/nr/results.htm (accessed February 1, 2006).

Nyden, Paul J. 2005. "Miners Arrested near Massey Plant." *Charleston Gazette,* February 25.

———. 2010. "Group Appealing Blair Mountain's Removal from Historic Register." *Charleston Gazette,* January 16.

O'Connor, James. 1994. "Is Sustainable Capitalism Possible?" In *Is Capitalism Sustainable? Political Economy and the Politics of Ecology,* edited by M. O'Connor, 152–75. New York: The Guilford Press.

OHVEC (Ohio Valley Environmental Coalition). 2002. "Massey Valley Fill Disaster, Lyburn, WV, July 19." Ohio Valley Environmental Coalition. http://www.ohvec.org/galleries/mountaintop_removal/006/index.html (accessed April 6, 2009).

———. 2004. "Coalfield Residents Speak the TRUTH." *Winds of Change: the Newsletter of the Ohio Valley Environmental Coalition,* July. http://www.ohvec.org/newsletters/woc_2004_07/article_18.html (accessed March 20, 2005).

———. [2007]. "Truck Facts the Industry Is Trying to Ignore." Ohio Valley Environmental Coalition. http://www.ohvec.org/issues/overweight_coal_trucks/sb583_fact.html (accessed May 30, 2007).

Omi, Michael, and Howard Winant. 1994. *Racial Formation in the United States from the 1960s to the 1990s.* New York: Routledge.

———. 2009. "Thinking through Race and Racism." *Contemporary Sociology: A Journal of Reviews* 38 (2): 121–25.

Ortner, Sherry B. 1972. "Is Female to Male as Nature Is to Culture?" *Feminist Studies* 1 (2): 5–31.

Pancake, Ann. 2007. *As Strange as This Weather Has Been.* Berkeley, Calif.: Shoemaker & Hoard.

Parsons, Brinckerhoff, Quade & Douglas, Inc. 2001. NCHA Strategic Management Action Plan. http://www.coalheritage.org/ (accessed January 26, 2006).

Pascale, Celine-Marie. 2007. *Making Sense of Race, Class, and Gender: Commonsense, Power, and Privilege in the United States.* New York: Routledge.

Pascoe, C. J. 2007. *Dude, You're a Fag: Masculinity and Sexuality in High School.* Berkeley: University of California Press.

Pateman, Carole. 1989. *The Disorder of Women: Democracy, Feminism and Political Theory.* Stanford, Calif.: Stanford University Press.

Perry, Pamela. 2002. *Shades of White: White Kids and Racial Identities in High School.* Durham, N.C.: Duke University Press.

Plumwood, Val. 1993. *Feminism and the Mastery of Nature.* New York: Routledge.

Polanyi, Karl. 1944. *The Great Transformation: The Political and Economic Origins of Our Times.* Boston: Beacon Press.

Porterfield, Mannix. 2009. "Wind Power OK with Randolph Senator If Rules Are Specific." *Beckley (W.Va.) Register-Herald,* February 13.

Precourt, Walter. 1983. "The Image of Appalachian Poverty." In *Appalachia and America,* edited by A. Batteau, 86–110. Lexington: University Press of Kentucky.

Probyn, Elsbeth. 1998. "Mc-Identities: Food and the Familial Citizen." *Theory, Culture, & Society* 15 (2): 155–73.

———. 1999. "Bloody Metaphors and Other Allegories of the Ordinary." In *Between Woman and Nation: Nationalisms, Transnational Feminisms, and the State,* edited by C. Kaplan, N. Alarcón and M. Moallem, 47–62. Durham, N.C.: Duke University Press.

Pudup, Mary Beth. 1990. "Women's Work in the West Virginia Economy." *West Virginia History* 49:7–20.

Pudup, Mary Beth, Dwight B. Billings, and Altina Waller. 1995. *Appalachia in the Making: The Mountain South in the Nineteenth Century.* Chapel Hill: University of North Carolina Press.

Rand, Ayn. 1957. *Atlas Shrugged.* New York: Random House.

Ratcliffe, Krista. 2005. *Rhetorical Listening: Identification, Gender, Whiteness.* Carbondale: Southern Illinois University Press.

Revkin, Andrew. 2009. "Coal Industry Wins a Round on Mining." *New York Times,* February 13.

Risman, Barbara. 2009. "From Doing to Undoing: Gender as We Know It." *Gender and Society* 23 (2): 81–84.

Rocheleau, Dianne E., Barbara P. Thomas-Slayter, and Esther Wangari. 1996. *Feminist Political Ecology: Global Issues and Local Experiences, International Studies of Women and Place.* New York: Routledge.

Roddy, Dennis. 2006. "Federal Officials Long Saw Problems at Sago Mine." *Pittsburgh Post Gazette,* January 19.

Roediger, David. 1991. *Wages of Whiteness: Race and the Making of the American Working Class.* New York: Verso.

Rosaldo, Renato. 1993. *Culture and Truth: The Remaking of Social Analysis.* Boston: Beacon Press.

Ross, Andrew. 2000. *The Celebration Chronicles: Life, Liberty, and the Pursuit of Property Values in Disney's New Town.* London: Verso.

Rowbotham, Sheila. 1974. *Women's Consciousness, Man's World.* London: Pelican.

Rubin, Gayle. 1997. "The Traffic in Women: Notes on the 'Political Economy' of Sex." In *The Second Wave: a Reader in Feminist Theory,* edited by L. Nicholson, 27–62. New York: Routledge.

Said, Edward W. 1978. *Orientalism.* New York: Random House.

Salleh, Ariel. 1997. *Ecofeminism as Politics: Nature, Marx, and the Postmodern.* New York: St. Martin's Press.

Sandoval, Chela. 2000. *Methodology of the Oppressed*. Minneapolis: University of Minnesota Press.

Saugères, Lise. 2002. "Of Tractors and Men: Masculinity, Technology, and Power in a French Farming Community." *Sociologia Ruralis* 42 (2): 143–59.

Savage, Lon. 1990. *Thunder in the Mountains: The West Virginia Mine War, 1920–21*. Pittsburgh: University of Pittsburgh Press.

Schnayerson, Michael. 2008. *Coal River: How a Few Brave Americans Took on a Powerful Company—and the Federal Government—to Save the Land They Love*. New York: Farrar, Straus and Giroux.

Scott, James. 1998. *Seeing Like a State: How Certain Schemes to Improve the Human Condition Have Failed*. New Haven, Conn.: Yale University Press.

Scott, Joan W. 1988. "Deconstructing Equality-Versus-Difference, or The Uses of Poststructuralist Theory for Feminism." *Feminist Studies* 14 (1): 32–50.

Scott, Rebecca. 2001. "Masculinity, Political Culture, and Community in Mountaintop Removal Discourse." Master's thesis, sociology, University of California, Santa Cruz.

Semple, Ellen Churchill. 1901 (1995). "The Anglo-Saxons of the Kentucky Mountains." In *Appalachian Images in Folk and Popular Culture*, edited by W. K. McNeil, 145-–74. Knoxville: University of Tennessee Press.

Sherman, Jerome L. 2006. "Mexico's Mine Crisis: 'We Want Our Loved Ones.'" *Pittsburgh Post-Gazette*, September 10.

Shiva, Vandana. 2005. *Earth Democracy: Justice, Sustainability and Peace*. Cambridge, Mass.: South End Press.

Shogan, Robert. 2004. *The Battle of Blair Mountain: The Story of America's Largest Labor Uprising*. Boulder, Colo.: Westview Press.

Simpson, Cameron. 2004. "Pulled on a Leash by a Tormentor from West Virginia Trailer Park." *Herald* (Glasgow), May 7.

Slack, Jennifer Daryl. 1996. "The Theory and Method of Articulation in Cultural Studies." In *Stuart Hall: Critical Dialogues in Cultural Studies*, edited by D. Morley and K.-H. Chen, 112–27. New York: Routledge.

Sledge, Colby, and Anne Paine. 2008. "Officials Test TVA Sludge." *Tennessean*, December 24.

Sloan, Bob. 2005. "Coal Industry's Models of Success Are Just False Fronts." *Lexington Herald-Leader*, August 8.

Slotkin, Richard. 1973 (2000). *Regeneration through Violence: The Mythology of the American Frontier*. Norman: University of Oklahoma Press.

Sludge Safety Project. [2009]. "Marsh Fork Elementary, Massey Energy's Shumate Coal Sludge Impoundment and Goals Coal Prep Plant." http://www.sludgesafety.org/what_me_worry/marsh_fork/index.html (accessed March 19, 2009).

Smith, Andrea. 2005. *Conquest: Sexual Violence and American Indian Genocide.* Cambridge, Mass.: South End Press.

———. 2008. *Native Americans and the Christian Right: The Gendered Politics of Unlikely Alliances.* Durham, N.C.: Duke University Press.

Smith, Linda Tuhiwai. 2001. *Decolonizing Methodologies: Research and Indigenous Peoples.* London: Zed Books.

Smith, Neil. 1992. "Contours of a Spatialized Politics: Homeless Vehicles and the Production of Geographical Scale." *Social Text* 33:54–81.

———. 2004. "Scale Bending and the Fate of the National." In *Scale and Geographic Inquiry,* edited by E. Sheppard and E. B. McMaster, 192–212. Malden, Mass.: Blackwell.

Somers, Margaret R. 1995. "Narrating and Naturalizing Civil Society and Citizenship Theory: The Place of Political Culture and the Public Sphere." *Sociological Theory* 13 (3): 229–74.

Stallybrass, Peter. 1999. "Worn Worlds: Clothes, Mourning, and the Life of Things." In *Cultural Memory and the Construction of Identity,* edited by D. Ben-Amos and L. Weissberg, 27–44. Detroit: Wayne State University Press.

Steedman, Carolyn. 1986. *Landscape for a Good Woman: A Story of Two Lives.* New Brunswick, N.J.: Rutgers University Press.

Steele, Terry, Jim Foster, Charles Branham, and Charles Nelson. 2007. "Interests of the Working Man: Citizen Groups Are Working to SAVE the Mountain State." *Charleston Gazette,* June 7, Opinion.

Stein, Arlene. 2001. *The Stranger Next Door: The Story of a Small Community's Battle over Sex, Faith, and Civil Rights.* Boston: Beacon Press.

Stein, Rachel. 2004. Introduction to *New Perspectives in Environmental Justice: Gender, Sexuality and Activism,* edited by R. Stein, 1–17. New Brunswick, N.J.: Rutgers University Press.

Stewart, Kathleen. 1996. *A Space on the Side of the Road: Cultural Poetics in an "Other" America.* Princeton, N.J.: Princeton University Press.

———. 2007. *Ordinary Affects.* Durham, N.C.: Duke University Press.

Stewart, Susan. 1984. *On Longing: Narratives of the Miniature, the Gigantic, the Souvenir, the Collection.* Baltimore, Md.: Johns Hopkins University Press.

Stockman, Vivian. 2004. "Coalfield Flooding: We've Already Answered This Question, Again, and Again, and . . ." *Winds of Change: The Newsletter of the Ohio Valley Environmental Coalition,* July. http://www.ohvec.org/newsletters/ woc_2004_07/article_07.html (accessed November 8, 2005).

Stoler, Ann Laura. 1995. *Race and the Education of Desire: Foucault's History of Sexuality and the Colonial Order of Things.* Durham, N.C.: Duke University Press.

———. 2002. *Carnal Knowledge and Imperial Power.* Berkeley: University of California Press.

Stout, Ben M., and Jomana Papillo. 2004. "Well Water Quality in the Vicinity of a Coal Slurry Impoundment Near Williamson, West Virginia." Working paper. Wheeling, W. Va.: Wheeling Jesuit University.

Strauss, Julius. 2006. "Trapped Miners Rescued." *Toronto Globe and Mail,* January 30.

Sturken, Marita. 2004. "The Aesthetics of Absence: Rebuilding Ground Zero." *American Ethnologist* 31 (3): 311–25.

Tallichet, Suzanne E., Meredith M. Redlin, and Rosalind P. Harris. 2003. "What's a Woman to Do? Globalized Gender Inequality in Small-Scale Mining." In *The Socio-Economic Impacts of Artisanal and Small-Scale Mining in Developing Countries,* edited by G. M. Hilson, 205–19. Exton, Penn.: A. A. Balkerma.

Thomas, Evan. 2004. "Explaining Lynndie England: How Did a Wispy Tomboy Behave Like a Monster at Abu Ghraib?" *Newsweek,* May 15.

Thompson, E. P. 1975. *Whigs and Hunters: The Origin of the Black Act.* New York: Pantheon Books.

Thongchai, Winichakul. 1994. *Siam Mapped: A History of the Geo-body of a Nation.* Chiang Mai, Thailand: Silkworm Books.

Thrift, Nigel. 2008. *Non-representational Theory.* London: Routledge.

Tomes, Nancy. 1998. *The Gospel of Germs: Men, Women, and the Microbe in American Life.* Cambridge, Mass.: Harvard University Press.

Trail Scouts. [2005]. "What Are the Trail Scouts?" http://www.trailscouts.com/ ts-what.htm (accessed December 15, 2005).

Trails Heaven. [2005]. Hatfield–McCoy Trails Official Web Site Home Page. http://www.trailsheaven.com/ (accessed March 5, 2005).

Trotter, Joe William. 1990. *Coal, Class, and Color: Blacks in Southern West Virginia, 1915–1932.* Urbana: University of Illinois Press.

Tsing, Anna. 1993. *In the Realm of the Diamond Queen: Marginality in an Out-of-the-Way Place.* Princeton, N.J.: Princeton University Press.

———. 2005. *Friction: An Ethnography of Global Connection.* Princeton, N.J.: Princeton University Press.

Twine, France Winddance. 1997. "Brown-Skinned White Girls: Class, Culture, and the Construction of White Identity in Suburban Communities." In *Displacing Whiteness: Essays in Social and Cultural Criticism,* edited by R. Frankenberg, 214–43. Durham, N.C.: Duke University Press.

UCC (United Church of Christ). 1987. *Toxic Wastes and Race in the United States: A National Report on the Racial and Socio-Economic Characteristics of Communities with Hazardous Waste Sites.* New York: United Church of Christ Commission on Racial Justice.

U.S. Census 2000. Online database. http://www.census.gov/.

U.S. Department of Labor. [2005]. "Table 5. Union Affiliation of Employed Wage and Salary Workers by State." Bureau of Labor Statistics. http://www.bls .gov/news.release/union2.to5.htm (accessed November 5, 2005).

————. [2006]. "Table 3. Union Affiliation of Employed Wage and Salary Workers by Occupation and Industry." Bureau of Labor Statistics. http://www.bls.gov/news.release/union2.t03.htm (accessed February 3, 2006).

————. [2009]. "Table 18. Employed Persons by Detailed Industry, Sex, Race, and Hispanic or Latino Ethnicity." Bureau of Labor Statistics. http://www.bls.gov/cps/cpsaat18.pdf (accessed May 5, 2010).

U.S. Senate. 2002. Subcommittee on Clean Air, Wetlands, and Climate Change, Committee on Environment and Public Works. *Clean Water Act: Review of Proposed Revisions to Section 404 Definitions of "Fill" and "Dredged Fill."* 107th Cong., 2nd Sess. June 6.

Valocchi, S. 1994. "The Racial Basis of Capitalism and the State, and the Impact of the New Deal on African Americans." *Social Problems* 41 (3): 347–62.

Visweswaran, Kamala. 1994. *Fictions of Feminist Ethnography*. Minneapolis: University of Minnesota Press.

Vollers, Maryanne. 1999. "Razing Appalachia." *Mother Jones*, July/August.

Wald, Matthew L. 2007. "Cleaner Coal Is Attracting Some Doubts." *New York Times*, February 21.

————. 2008. "Mounting Costs Slow the Push for Clean Coal." *New York Times*, May 30.

Walker, David. 1829 (1995). *David Walker's Appeal in Four Articles, Together with a Preamble, to the Coloured Citizens of the World, but in Particular and Very Expressly, To Those of the United States of America*. 1829. Reprinted. New York: Hill and Wang.

Wall, Glenda, and Stephanie Arnold. 2007. "How Involved Is Involved Fathering?" *Gender and Society* 21 (4): 508–27.

Waller, Altina. 1988. *Feud: Hatfields, McCoys, and Social Change*. Chapel Hill: University of North Carolina Press.

Ward, Ken, Jr. 1998. "400 Arch Miners in Logan Face Layoff Threat." *Charleston Gazette*, October 31.

————. 2001. "Massey's Record Near the Worst, Inspectors Say." *Charleston Gazette*, December 7.

————. "Bush Administration Plan Broadens Valley Fill Rule Changes." *Charleston Gazette*, April 26.

————. 2005. "Feds End Criminal Probe of Slurry Spill." *Charleston Gazette*, May 11.

————. 2006a. "Report Criticizes Gap in Sago Rescue Info." *Charleston Gazette*, July 25.

————. 2006b. "Disasters Get the Headlines, but Most Miners Killed on the Job Die Alone: Ninety Percent of Fatalities Caused by Violating Existing Laws." *Charleston Gazette*, November 5.

————. 2006c. "Mine Bill Targets MSHA." *Charleston Gazette*, February 2.

———. 2007. "MSHA Citations Detail Sago Problems." *Charleston Gazette*, June 3.

———. 2008a. "Candidates Agree on Mountaintop Removal." *Charleston Gazette*, September 21.

———. 2008b. "Coal River Mine Permit Challenged by Wind Proponents." *Charleston Gazette*, December 19.

———. 2008c. "Mine's Selenium Deforms Fish, Expert Says." *Charleston Gazette*, April 27.

———. 2008d. "MSHA Violations on Rise at Alma No. 1." *Charleston Gazette*, May 20.

———. 2008e. "UMW Taking Up Mountaintop Fight?" *Charleston Gazette*, April 6.

———. 2009a. "Blair Mountain Added to National Register." *Charleston Gazette*, March 30.

———. 2009b. "Goodwin Blocks Corps from Issuing Streamlined Mountaintop Removal Permits." *Charleston Gazette*, March 31.

———. 2009c. "Massey Protest Update: 14 Arrested, Accusations Fly." *Coal Tattoo: Mining's Mark on Our World*, June 18. http://blogs.wvgazette.com/coaltattoo/2009/06/18/massey-protest-update-14-arrested-accusations-fly/#more-813 (accessed June 18, 2009).

———. 2009d. "State Moves to Delist Blair Mountain from Historic Register." *Charleston Gazette*, April 6.

———. 2010. "Mountaintop Removal Mining: EPA Study Confirms Damage." *Charleston Gazette*, April 4.

Ware, Vron. 1992. *Beyond the Pale: White Women, Racism, and History*. London: Verso.

Warren, Karen J. 1997. *Ecofeminism: Women, Culture, Nature*. Bloomington: Indiana University Press.

Warrick, Joby. 2004. "Appalachia Is Paying Price for White House Rule Change." *Washington Post*, August 17.

Watts, Michael. 2001. "Petro-Violence: Community, Extraction, and Political Ecology of a Mythic Commodity." In *Violent Environments*, edited by N. L. Peluso and M. Watts, 189–212. Ithaca, N.Y.: Cornell University Press.

Weber, Max. 1958. *The Protestant Ethic and the Spirit of Capitalism*. New York: Charles Scribner's Sons.

Weissberg, Liliane. 1999. "Memory Confined." In *Cultural Memory and the Construction of Identity*, edited by D. Ben-Amos and L. Weissberg, 45–76. Detroit: Wayne State University Press.

Weller, Jack E. 1965 (1995). *Yesterday's People: Life in Contemporary Appalachia*. Lexington: University Press of Kentucky.

Wellman, David. 2007. *Not Quite White: White Trash and the Boundaries of Whiteness*, by Matt Wray. *Contemporary Sociology: A Journal of Reviews* 36 (6): 556–57.

West, Candace, and Don H. Zimmerman. 1987. "Doing Gender." *Gender and Society* 1(2): 125–51.

West Virginia Archives & History. [2006]. "'Devil Anse Hatfield and the Hatfield–McCoy Feud." West Virginia Division of Culture and History. http://www.wvculture.org/history/notewv/hatfield.html (accessed May 30, 2006).

West Virginia Coal Association. 2006. "Coal Facts 2006." http://www.wvcoal.com/news/news-archive/30-west-virginia-coal-association-releases-2006-coal-facts.html (accessed March 3, 2007).

———. 2008. "Coal Facts 2008." http://www.wvcoal.com/resources/coal-facts.html (accessed January 9, 2009).

West Virginia Legislature's Office of Reference and Information. 2005. "Joint Commission on Economic Development." *Interim Highlights* VII(I) July: 6.

West Virginia Office of Miners' Health, Safety and Training. 2009. "Summary of Fatal Mining Accidents 1997–2009." http://www.wvminesafety.org/fatal97.htm (accessed January 28, 2009).

Wiegman, Robin. 1999. "Whiteness Studies and the Paradox of Particularity." *boundary 2* 26 (3): 115–50.

———. 2003. "Intimate Publics: Race, Property, and Personhood." In *Race, Nature, and the Politics of Difference*, edited by D. S. Moore, J. Kosek and A. Pandian, 296–319. Durham, N.C.: Duke University Press.

Wigginton, Eliot. 1972. *The Foxfire Book: Hog Dressing, Log Cabin Building, Mountain Crafts and Foods, Planting by the Signs, Snake Lore, Hunting Tales, Faith Healing, Moonshining, and Other Affairs of Plain Living.* Garden City, N.J.: Doubleday.

Wilen, John. 2009. "World Coal Market Hits Home." *Charleston Gazette,* April 29.

Williams, John Alexander. 2002. *Appalachia: A History.* Chapel Hill: University of North Carolina Press.

Williams, Patricia. 1991. *The Alchemy of Race and Rights: Diary of a Law Professor.* Cambridge, Mass.: Harvard University Press.

Williams, Raymond. 1973. *The Country and the City.* New York: Oxford University Press.

———. 1977. *Marxism and Literature.* Oxford: Oxford University Press.

Willott, Sara, and Christine Griffin. 1997. "'Wham Bam, Am I a Man?' Unemployed Men Talk about Masculinities." *Feminism & Psychology* 7 (1): 107–28.

Wilson, Darlene, and Patricia Beaver. 1999. "Transgressions in Race and Place: The Ubiquitous Native Grandmother in America's Cultural Memory." In *Neither Separate nor Equal: Women, Race, and Class in the South,* edited by B. Smith, 34–56. Philadelphia: Temple University Press.

Wilson, Robert F. 1974. "*Deliverance* from Novel to Film: Where Is Our Hero?" *Literature/Film Quarterly* 2 (1): 52–58.

Workforce West Virginia. [2005]. *West Virginia Labor Force Estimates* http://

www.wvbep.org/bep/lmi/datarel/DRLMI134.HTM (accessed November 5, 2005).

Wray, Matt. 2006. *Not Quite White: White Trash and the Boundaries of Whiteness.* Durham, N.C.: Duke University Press.

WVInc. 2005. "On the Record with Don Blankenship." *WVInc Online,* August 1. http://www.wvinconline.com/news/news/143.aspx (accessed January 12, 2006).

Zukin, Sharon. 1991. *Landscapes of Power: From Detroit to Disney World.* Berkeley: University of California Press.

Index

REBECCA R. SCOTT is assistant professor of sociology at the University of Missouri.